THE COMMONWEALTH AND INTERNATIONAL LIBRARY
Joint Chairmen of the Honorary Editorial Advisory Board
SIR ROBERT ROBINSON, O.M., F.R.S., LONDON
DEAN ATHELSTAN SPILHAUS, MINNESOTA

GEOPHYSICS DIVISION
General Editors: J. A. JACOBS AND J. T. WILSON

THE EARTH'S AGE
AND
GEOCHRONOLOGY

THE EARTH'S AGE
AND
GEOCHRONOLOGY

DEREK YORK

AND

RONALD M. FARQUHAR

Geophysics Division, Department of Physics,
University of Toronto, Toronto, Canada

PERGAMON PRESS

OXFORD · NEW YORK · TORONTO
SYDNEY · BRAUNSCHWEIG

COLLEGE OF THE SEQUOIAS
LIBRARY

Pergamon Press Ltd., Headington Hill Hall, Oxford

Pergamon Press Inc., Maxwell House, Fairview Park, Elmsford,
New York 10523

Pergamon of Canada Ltd., 207 Queen's Quay West, Toronto 1

Pergamon Press (Aust.) Pty. Ltd., 19a Boundary Street,
Rushcutters Bay, N.S.W. 2011, Australia

Vieweg & Sohn GmbH, Burgplatz 1, Braunschweig

Copyright © 1972 Pergamon Press Ltd.

*All Rights Reserved. No part of this publication may be
reproduced, stored in a retrieval system, or transmitted, in any
form or by any means, electronic, mechanical, photocopying,
recording or otherwise, without the prior permission of
Pergamon Press Ltd.*

First edition 1972

Library of Congress Catalog Card No. 72–157001

Printed in Great Britain by A. Wheaton & Co., Exeter

This book is sold subject to the condition
that it shall not, by way of trade, be lent,
resold, hired out, or otherwise disposed
of without the publisher's consent,
in any form of binding or cover
other than that in which
it is published.

08 016387 4

CONTENTS

CONTENTS

PREFACE

AFTER a bright beginning, early in the present century, geochronology developed slowly. The late 1940s and early 1950s, however, saw an enormous burst of activity which clearly established the feasibility and great promise of the K–Ar, Rb–Sr and U,Th–Pb methods. The years since then have seen significant improvements in experimental method and even more remarkable progress in interpretation of radiometric age data. This rapid development stands in marked contrast with the dearth of books on geochronology. The authors have endeavoured to ameliorate this situation by writing an outline of modern geochronological methods, interpretations and some applications. In a book of this size a considerable degree of selection must be exercised. By far the major part of the book's emphasis has been placed on the K–Ar, Rb–Sr and U,Th–Pb methods which form the backbone of geochronology. The relatively new fossil fission track method of dating is described briefly but with sufficient detail to prepare the student to read the research literature. One chapter is devoted to the subject of lead isotopes.

After an introductory first chapter, Chapters 2–7 are concerned with the methodology of dating. For the sake of clarity the authors have first (in Chapter 4) discussed the application of the experimental technique to idealized, undisturbed systems. In Chapter 5 more realistic situations are dealt with where significant daughter isotope diffusion has occurred. The importance of long cooling histories is discussed in Chapter 7. The concept that in plutonic environments daughter isotope retention may often not commence until long after crystallization, or the peak of metamorphism, has only since 1960 received widespread consideration. This idea has been most noticeable in European and British investigations, but North American workers are beginning to give it attention.

Chapters 8–12 are concerned with the applications of geochronology and here the effects of selectivity will be particularly evident. An attempt has been made to present applications of widespread interest. Clearly a discussion of the Phanerozoic Time-scale must find a place in a book such as this, while the question of the pulse of the earth and the establishment of chronology of the reversals of polarity of the earth's magnetic field are surely topics of great interest to earth scientists as a whole.

The question of the age of the earth is discussed at some length. Inevitably this entails a review of the dating of meteorites, as these objects are almost always intimately involved in calculations of the earth's age. In view of the non-terrestrial origin of meteorites, and having regard to the considerable uncertainties regarding their real origins and histories, it is somewhat unsatisfactory that they should play such a central role in calculations of the age of the earth. That they do, however, is an inescapable fact. As is shown in Chapter 12, several calculations indicate that the earth and the meteorites shared the same lead isotopic composition approximately $4 \cdot 6$ billion years ago (i.e. $4 \cdot 6 \times 10^9$ years ago) and it is usually considered that this value may be taken as representative of the age of the earth. I–Xe dating suggests that the chemical elements constituting the meteorites and probably the earth were synthesized less than 100 million years prior to the formation of these bodies.

It is hoped that the book will be helpful to students at the undergraduate and graduate levels, and to geologists and geophysicists. Geologists wishing to make use of geochronological methods in their studies should readily be able to assess the methods most useful to them and numerous references are given as a guide to further reading.

The authors are grateful to Drs. C. T. Harper and R. L. Grasty for reading the manuscript and making useful comments. Mrs Lydia York considerably speeded up final production of the book by typing the first draft while Mrs Debnam skilfully typed the final manuscript. One of us (D.Y.) is indebted to the National Research Council of Canada for an award of a Senior Research Fellowship which enabled him to spend a sabbatical leave working on the manuscript at the department of Geophysics and Geochemistry, The Australian National University, Canberra, Australia. The hospitality and facilities offered by Professor J. C. Jaeger, Drs. W. Compston, I. McDougall and J. R. Richards greatly aided the completion of this work.

CHAPTER 1

INTRODUCTION

To the average person, living in modern society, the measurement of time is accepted as a matter of fact. It is not until we are without a time-piece of some sort that we realize the extent to which we depend on accurate time-keeping. In the sciences the demand for improved methods of time-keeping has led to remarkable technical advances in this field. Nuclear physicists, for example, in their studies of fundamental particles, have had to examine processes and reactions which occur within intervals as short as 10^{-9} second. Astronomers at widely separated observatories have recently used atomic clocks synchronized to better than 10^{-6} second to study the mysterious radio stars known as quasars. At the other end of the scale, geologists and cosmologists have desired methods for dating events which occurred as long as 10^9 to 10^{10} years ago, in the history of the earth and the universe.

The nuclear physicist and the astronomer, of course, use very different sorts of mechanism for their chronometers than those employed by the geologist and the cosmologist. But although the technical problems of time measurement are consequently very different, the basic properties and requirements of the devices used are identical. An accurate chronometer for measuring time intervals must contain some sort of mechanism which operates at a known or predictable (but not necessarily constant) rate. This mechanism must be linked to a recording system of some kind, in which *events* marking the beginning and end of the time interval must be clearly and sharply recorded. These events must have no effect on the rate of action of the mechanism, and the recording system must be *selective* in the sense that other events which may occur in the time interval are either not recorded, or do not alter the recorded features following the initial event.

A chronometer which fails to meet these requirements will give incorrect or uncertain results. This does not mean that the results are useless—in

1

an area in which no previous measurements are available a piece of data providing even an estimate of the order of magnitude is certainly useful. What is most important is that some estimate of the magnitude of the uncertainty be made so that the data may not be overrated when it is removed from its original context.

In making measurements of time intervals the physicist has some measure of control over the operation of his chronometer, and its recording system. The geologist, on the other hand, must depend on natural terrestrial processes and systems to provide the mechanism and the recording facilities of his device. He can determine the instant of the final event, but he has no control over the initial event. This in fact is probably the most serious limitation because it places a severe restriction on the type of geological event which can be dated, and hence limits the number of useful dates which can be obtained.

Until the beginning of the present century geologists had no really satisfactory mechanisms for measuring time or time intervals. But the span of geological time was not entirely unknown, and several attempts were in fact made to gain some idea of its length, even though the methods used were admittedly imperfect. It is worth briefly reviewing two of these because they point up so clearly the requirements of methods of accurate time measurements in geological systems.

One of the earliest was an estimate of the age of the oceans. The oceans contain salt, conveyed to them by rivers which drain the continental land masses. The salt is dissolved in the river water in the course of the erosion of surface rocks. Joly (1899) noted that, if one knew the total amount of salt in the oceans at the present time and the rate at which salt was being delivered to the oceans, the "age" of the oceans (or more specifically, the time when the oceans contained no salt) could be estimated.

Many questions concerning this estimate will immediately occur to the reader. Is the present rate of supply of salt to the oceans representative of past rates? How much of the salt now in the oceans has been recycled, that is, deposited in sediments and subsequently re-eroded to return again to the oceans? Are there other processes by which this recycling can take place? These questions are not trivial, and must be answered before the worth of any age estimate can be assessed. In fact the problem soon assumes the proportions of a major inquiry into the geochemical relations between the oceans and the continents. As knowledge of these relations has grown, the estimates of age have increased; so, unfortunately, have the numerical uncertainties in these estimates. Joly (1899) computed a value of 90 million years; Livingstone (1963) estimates that the figure could lie between 1300 million years and 2500 million years.

The failure of this method to yield a really useful age estimate lies firstly in the difficulty of determining the rate of the chronometric process (the rate of salt delivery) and secondly in the imperfect nature of the recording system (the ocean). For similar reasons, the estimate by Kelvin (1899) of the age of the earth proved erroneous. This estimate was obtained by computing the time for the earth to cool by conduction from a supposedly molten state to its present condition of low surface heat flow. Kelvin originally believed that the only major question in this method was the value of the initial temperature, the "initial event" being dated. The age estimates ranged from the 20 million year figure which Kelvin preferred, to at most a few hundred million years. These values were in disagreement with other estimates of the duration of terrestrial processes, such as ages based on rates of evolution of various fauna, made by the geologists. These discrepancies were resolved (in favour of the geologists) by the discovery of radioactivity. Radioactive elements decaying within the earth act as sources of heat, and materially prolong the cooling process. Recently the mechanism of heat transfer within the earth has been shown to be much more complex than the simple conductive type envisaged by Kelvin. As with the age of the oceans, the original problem has reverted to a more thorough study of the processes involved.

This development represents in fact the real value of these early age estimates. Because of the uncertainties and incongruities in the results, investigators were led to consider the processes in more detail, and to undertake further measurements. Curiously enough, the discovery of radioactivity, which disqualified Kelvin's estimate of the age of the earth, provided the mechanism for making the most accurate estimates of the ages of geological events. In the following chapters we will examine the decay process in some detail, and note to what extent minerals and rocks in which this process is going on meet the requirements of good chronometric systems.

CHAPTER 2

RADIOACTIVITY

2.1. Radioactive Decay

The phenomenon of radioactivity was discovered in 1896 by the French physicist Henri Becquerel. The many investigators who subsequently took up the study of this phenomenon showed that it was a manifestation of the breakdown or "decay" of unstable atoms. This process is a spontaneous one, and isotopes which undergo decay become nuclei of different elements as a consequence.

The vast majority of naturally occurring elements have isotopes with stable nuclei. However, there do exist several with *unstable* nuclei. For our purposes, the most important of these are K^{40}, Rb^{87}, U^{238}, U^{235}, Th^{232} and Re^{187}. The great potentiality of such unstable nuclei as geological clocks was recognized soon after the discovery of radioactivity, and several geochronological studies were attempted during the first decade of this century.

The decay process is an exothermic one, in which the excess energies of the unstable nuclei are usually carried away through the emission of atomic alpha or beta particles, and gamma radiation. Because the nature of the emission determines the new nucleus formed, it is worth while examining the properties of these decay processes in more detail.

α-DECAY

An α-particle is the nucleus of a He^4 atom. It is a close grouping of two protons and two neutrons. Accordingly, its electrical charge is two units and its mass is four units. This particle is ejected from an unstable nucleus during the α-decay. After such an event the residual, or daughter nucleus now contains two fewer protons than the original, unstable parent nucleus. It is thus the nucleus of a new element whose nuclear charge is two units less than that of the parent and whose nuclear mass is four units lower, e.g.

$$_{92}U^{238} \rightarrow {}_{90}Th^{234} + \alpha.$$

β-DECAY

Some radioactive nuclei decay by the emission of negative electrons from the unstable parent nucleus. The charge on the nucleus effectively goes up one unit, while the nuclear mass is essentially unchanged. Under the heading β-decay we can also include radioactive decay by electron capture. In this mode of disintegration an orbital electron is captured by the nucleus—the nuclear charge now going down one unit with no significant change in mass. One of the most important naturally occurring radioactivities, that of K^{40}, takes this form:

$$K^{40} + \text{orbital electron} \rightarrow Ar^{40} + \gamma.$$

The radioactivity of K^{40} is also unusual because the ordinary β-decay process competes with the electron capture process. So we may also spontaneously have:

$$K^{40} \rightarrow Ca^{40} + \beta^-.$$

γ-EMISSION

This does not occur as an independent form of radioactivity but as a concomitant of α- or β-decay. γ-rays are high energy electro-magnetic radiation emitted by an excited nucleus as it drops into a less excited state.

Radioactive decay was shown by Rutherford (1900) to follow an exponential law. The fundamental equation describing it may be written

$$\frac{dP}{dt} = -P\lambda, \tag{2.1}$$

where P = the number of parent atoms present at time t. λ is a constant characteristic of the particular radioactive element concerned. It is called the disintegration constant or decay constant.

Integration of this equation gives us

$$t = \frac{1}{\lambda} \ln \left\{ 1 + \frac{D}{P} \right\}, \tag{2.2}$$

where D = initial number of parent atoms − number of parent atoms after time t; P = number of parent atoms after time t.

Equation (2.2) is the fundamental equation of geochronology. It is extremely important to note that it was derived assuming that λ is a constant and that the only alteration in amount of daughter or parent in the system is due to radioactive decay. Neither assumption is trivial. That the usual forms of radioactivity carry on at a constant rate entirely unaffected by changes in the environment of the decaying atoms has long been accepted. However, in this connection it is interesting to note two points. One is that the K-capture process decay rate *is* altered by a change in external parameters such as pressure or state of chemical combination of the atom. The effect has been detected experimentally in Be^7 (Leininger *et al.*, 1951; Kraushaar *et al.*, 1953). Very fortunately, the K^{40} electron capture process seems to be much more resistant to change and the decay constant for this decay may be safely considered to be unaltered by pressure, etc. The second point concerns the possibility that the various "constants of nature" (i.e. the gravitational constant G, the radioactive decay constants, etc.) may be functions of time. This matter has been considered by Dirac (1939), Dicke (1959) and Kanasewich and Savage (1963), among others. It would seem that even if the "constants" have been varying, the amount of variation would not be such as to produce measurable discrepancies in geological ages based on the tenet of constant λ's. Accordingly disintegration constants will be treated throughout the remainder of the book as constants.

TABLE 2.1. VALUES OF HALF-LIVES AND DECAY
CONSTANTS IN CURRENT USE

Isotope	$\lambda(10^{-10} \text{ y}^{-1})$	$t_{\frac{1}{2}}(10^9 \text{ y})$
K^{40}	$\lambda e = 0 \cdot 585$ $\lambda \beta = 4 \cdot 72$ $\lambda = 5 \cdot 31$	$11 \cdot 8$ $1 \cdot 47$ $1 \cdot 31$
Rb^{87}	$0 \cdot 147$ or $0 \cdot 139$	$47 \cdot 0$ or $50 \cdot 0$
U^{238}	$1 \cdot 54$	$4 \cdot 51$
U^{235}	$9 \cdot 72$	$0 \cdot 713$
Th^{232}	$0 \cdot 499$	$13 \cdot 9$
Re^{187}	$0 \cdot 161$	$43 \cdot 0$

The respective uncertainties in these values are discussed in the text.

TABLE 2.2. BROAD ESTIMATES OF CONCENTRATIONS OF RADIOACTIVE AND "COMMON DAUGHTER" ELEMENTS IN COMMON ROCKS

	U (ppm)	Th (ppm)	Pb (ppm)	K %	Rb (ppm)	Sr (ppm)
Granite	4	15	20	3·5	200	300
Basalt	1	3	4	0·75	30	470
Ultramafic	0·02	0·08	0·1	0·004	0·5	50
Shale	4	12	20	2·7	140	300

The other assumption, that no gain or loss of parent or daughter other than by radioactive decay has occurred, is much more questionable and is one which has to be borne in mind at all times. Discussions on whether or not certain systems have remained closed during their history and on how one endeavours to produce unequivocal answers to such problems are given in Chapters 4, 5 and 7.

2.2. Criteria for Useful Radioactivities

For the purposes of geochronometry of the type we are considering, three major criteria have to be satisfied by a radioactive isotope. (a) The half-life of the radioactivity must be roughly of the order of the age of the earth (4·5 b.y.). (b) The isotope under consideration must be reasonably abundant in terrestrial rocks. (c) Significant enrichments of the daughter must occur. Table 2.1 shows the half-lives of the currently used isotopes. The abundances of these elements in some common rock types are seen in Table 2.2.

It becomes immediately apparent from Table 2.2 that the vast majority of published age data pertain to acid rocks simply because they have the greatest concentration of the appropriate radioactivities. As techniques have been improved figures have begun to appear for basic and even ultra-basic rocks.

2.3. Half-lives of U^{238}, U^{235}, Th^{232}, K^{40} and Rb^{87}

Physical determinations of decay constants are usually made by measuring the rate of decay dP/dt of a known quantity P of the radioactive isotope in question. As equation (2.1) shows, λ is then

$$\frac{dP/dt}{P},$$

the "specific activity" of the sample. These measurements are technically difficult (particularly where accuracies better than 1% are required), a fact that is reflected by the spread in values often obtained by different investigators.

U^{238}

This constant is usually considered to be the best known of the half-lives important in geochronometry. Most of the determinations have been made on samples of *natural* uranium by α-counting. Natural uranium also contains U^{235} and U^{234}, both of which are α-active. The U^{234} should contribute as much α-activity as the U^{238} so that the total activity minus that due to U^{235} must be divided by two to give the activity of U^{238}.

TABLE 2.3. MEASUREMENTS OF THE HALF-LIFE OF U^{238}
(After Aldrich and Wetherill)

	Specific activity natural uranium (d/min/mg)*	Specific activity U^{238} (d/min/mg)*	Half-life U^{238} (10^9 y)
Kovarik and Adams (1941)	1501 ± 3		4·50
Curtis, Stockman and Brown (1941)	1501 ± 3		4·50
Kienberger (1949)	1502 ± 2		4·50
Kienberger (1949)		$742·7 \pm 1·6$	4·49
Kovarik and Adams (1955)	1503 ± 3		4·51

* d/min stands for disintegrations per minute.

Kienberger (1949) avoided the latter problem by combining a measurement of the specific α-activity of enriched U^{234} with isotopic determinations of the U^{234} content of natural uranium. Table 2.3 lists the values found by various workers. The small scatter is noteworthy and a half-life of $4·51 \times 10^9$ years is usually adopted.

U^{235}

The α-activity of this isotope in natural uranium is only about 2% of that due to U^{238} and U^{234}. To overcome this problem, Fleming *et al.* (1952) counted highly enriched U^{235} samples in which the U^{238} activity was negligible and the U^{234} contributed 35% of the total α-activity. The half-life value of $7 \cdot 13 \times 10^8$ years found by these workers is usually accepted in geochronometric studies. Table 2·4 shows the published measurements available. The agreement could be better and a redetermination of this constant is needed.

TABLE 2.4. MEASUREMENTS OF THE HALF-LIFE OF U^{235}
(After Aldrich and Wetherill)

	Specific activity pure U^{235} (d/min/mg)	Half-life (10^8 y)
Clark, Spencer-Palmer and Woodward (1944)		7·64*
Knight (1950)	4480 ± 140	7·53 ± 0·25
Knight (1950)		7·10*
Fleming, Ghiorso and Cunningham (1952)	4740 ± 100	7·13 ± 0·16
Sayag (1951)		6·82 ± 0·29†
Würger, Meyer and Huber (1957)		6·84 ± 0·15

* As recalculated in Fleming *et al.* (1952).
† As recalculated in Würger *et al.* (1957).

When U–Pb ages are being calculated from "concordia" plots (see Chapter 5), the uncertainty in this half-life becomes very significant. Wasserburg *et al.* (1962) pointed out that their calculated age of 1200 m.y. for Texas zircons would be reduced to 1100 m.y. if they adopted a half-life just 2·2% below the mean given by Fleming *et al.* (1952).

Rb^{87}

In perhaps the worst plight of all is the half-life of Rb^{87}. The fundamental difficulty is that of counting accurately the large number of low-energy β-particles emitted by Rb^{87}. While the maximum energy of the electrons is 275 keV, the average energy is only 44 keV and scattering and absorption effects are a major problem. In Fig. 2.1 we have plotted the various values of this half-life which have been found since 1931. There is a disconcertingly huge scatter among all the methods of measurement. Several interesting points may be made, however. (a) The arithmetic average of all the measured half-lives shown is $5 \cdot 5 \times 10^{10}$ years. (b) If we average only the counting measurements carried out since 1957 we obtain $5 \cdot 25 \times 10^{10}$ years. (c) Since 1954 a number of measurements have

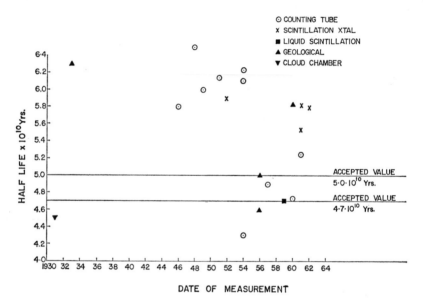

FIG. 2.1. Measured values of half-life of Rb^{87} as a function of the year of measurement.

indicated a value less than $5 \cdot 0 \times 10^{10}$ years. (d) Counting tube results dropped in value after 1954 by a significant amount. (e) Results found by crystal scintillation counting are systematically higher than those calculated from counting tube data and liquid scintillation experiments.

A recent measurement by McMullen *et al.* (1966) gives a half-life of $4 \cdot 72 \pm \cdot 04 \times 10^{10}$ years. In this interesting experiment about 1 kg of a Rb salt was freed of Sr and the Rb was allowed 7 years in which to generate Sr^{87}. The amounts of accrued Sr^{87} in 20-g aliquots of this sample were then measured by isotope dilution. In essence a miniature age measurement was carried out, only the age was known *a priori* and the half-life was treated as the unknown.

It will be apparent that the half-life of Rb^{87} is imperfectly known and a value may only currently be chosen arbitrarily. In recent years geochronologists have almost without exception calculated ages using a half-life equal to either $5 \cdot 0 \times 10^{10}$ years or $4 \cdot 7 \times 10^{10}$ years. The former value was given by Aldrich and Wetherill after comparison of Rb–Sr data with U–Pb figures. Such a method is, however, a makeshift one. $4 \cdot 7 \times 10^{10}$ years was found by Flynn and Glendenin (1959) using liquid scintillation counting, but it is not clear that this figure is preferable to those found by crystal scintillation counting. It may be significant, however, that this value is in agreement with the McMullen *et al.* result. In view of the uncertainties of the situation, much recalculation later on would be avoided and comparisons would be facilitated if one of these two values were adhered to until newer measurements converged to some generally acceptable figure. However, there is as yet no sign of any such convention being adopted. A given percentage uncertainty in the Rb^{87} half-life produces the same percentage uncertainty in a calculated age.

K^{40}

The decay scheme for K^{40} is shown in Fig. 2.2. Its dual decay is described by the two decay constants λe and $\lambda \beta$. The total decay constant equals $\lambda e + \lambda \beta$ and the so-called "branching ratio" R equals $\lambda e / \lambda \beta$.

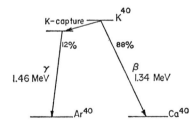

FIG. 2.2. Decay scheme of K^{40}.

Figure 2.3 shows the influence that errors in λe and $\lambda \beta$ have on a calculated age. It can be seen that, except in the case of meteorite ages, a given uncertainty in λe has more influence on the age than a similar percentage uncertainty in $\lambda \beta$. Determination of $\lambda \beta$ has usually been made by β-counting of thick sources of natural potassium because of the low specific activity.

Some thin source counting of sources enriched in K^{40} has been done but unfortunately the exact enrichment of K^{40} was not calculated. The fairly high maximum energy of the β-rays of $1\cdot34$ MeV and the smaller number of low-energy particles makes it easier to count the specific β-activity of K^{40} than that of Rb^{87}. A plot of experimentally determined values as a function of time of measurement is given in Fig. 2.4. The average of those measurements quoting uncertainties of less than ±2 disintegrations/g-sec is $27\cdot8$ disintegrations/g-sec. Aldrich and Wetherill (1958) recommended a value of $27\cdot6$ β/g-sec. This is usually used in geochronometric calculations

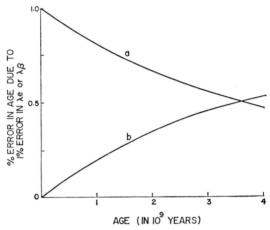

FIG. 2.3. Effect on K–Ar age of errors in $\lambda\beta$ and λe. $a = \lambda e$; $b = \lambda\beta$. (After Aldrich and Wetherill, 1958.)

and corresponds to a disintegration constant $\lambda\beta = 4\cdot72 \times 10^{-10}$ y^{-1}. While this figure must be considered to be uncertain by at least 2%, this will cause considerably less than 2% uncertainty in age.

λe is more difficult to measure. Customarily it is assumed that electron-capture takes place to the excited state of the Ar^{40} nucleus which then emits a $1\cdot46$-MeV γ-ray in reaching its ground state. Thus the specific γ-activity is found and used as a direct measure of λe. Measured values, as can be seen in Fig. 2.4, have fluctuated widely. The spread was ascribed by Wetherill mainly to the difficulty of determining the efficiency of the counting device used for γ-rays of the appropriate energy. Wetherill (1957) and McNair, Glover and Wilson (1956) used crystal scintillation counting to avoid this problem and obtained respectively the specific γ-activities $3 \cdot 39 \pm 0 \cdot 12$ γ/g-sec and $3 \cdot 33 \pm 0 \cdot 15$ γ/g-sec. In their review of the decay constants Aldrich and Wetherill (1958) suggested the value $3 \cdot 4$ γ/g-sec corresponding to $\lambda e = 0 \cdot 585 \times 10^{-10}$ y^{-1}. They considered this to have a $\pm 5\%$ uncertainty.

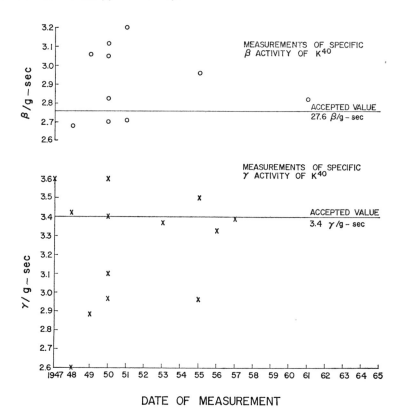

FIG. 2.4. Specific γ-activity and β-activity of K^{40} as a function of year of measurement.

Beyond this remains the possibility that some of the electron capture transitions occur to the ground state of Ar^{40} and are therefore not followed by γ-emission. Theory indicates that the ratio of the frequency of positron emission from K^{40} to the ground state of Ar^{40} to the frequency of

γ-emission is about 10^{-4}. Also, the ratio of electron capture to the ground state to positron emission is about 100. This would suggest that about 1 % of the electron captures are to the ground state of Ar^{40}. In an experiment to measure the positron-γ-ratio Tilley and Madansky (1959) set an upper limit of 2×10^{-4}, suggesting that fewer than 2 % of the electron captures go directly to the ground state.

Further determinations of λe and $\lambda \beta$ are keenly awaited. Meanwhile the recommended values of Aldrich and Wetherill, $\lambda e = 0 \cdot 585 \times 10^{-10}$ y^{-1} and $\lambda \beta = 4 \cdot 72 \times 10^{-10}$ y^{-1} are in general use.

Th^{232}

The three values extant for this constant are $t_{\frac{1}{2}} = 1 \cdot 39 \pm 0 \cdot 03 \times 10^{10}$ y, $1 \cdot 42 \pm 0 \cdot 07 \times 10^{10}$ y and $1 \cdot 39 \pm 0 \cdot 03 \times 10^{10}$ y. The first value was obtained by Kovarik and Adams (1938) by α-counting of natural thorium, it being assumed that equilibrium between Th^{232} and Th^{228} was established. Senftle *et al.* (1956) found the second figure using a NaI(Tl) scintillation spectrometer to count the $2 \cdot 62$ MeV γ's from Tl^{208}. Picciotto and Wilgain (1956) used a nuclear emulsion technique to get the third value above. The agreement among these workers is good and the value usually adopted is $\lambda = 4 \cdot 99 \times 10^{-11}$ y^{-1}.

CHAPTER 3

EXPERIMENTAL METHODS

3.1. Introduction

From the previous chapter we know that in a time t years, D atoms of a daughter element will have been formed from an original amount P_0 of parent, where

$$t = \frac{1}{\lambda} \log_e \left\{ 1 + \frac{D}{P} \right\} \tag{3.1}$$

and $P = P_0 - D =$ the number of parent atoms remaining at time t. The whole of isotopic geological age determination is based upon (3.1). This equation makes it clear that the method is really one of refined analytical chemical analysis, i.e. to obtain an "age" we have to ascertain the amounts of the parent and daughter *isotopes* present in the sample being investigated. We will now examine in some detail the practical considerations involved.

3.2. Mass Spectrometry

The mass spectrometer resulted from the "positive ray" studies of J. J. Thomson early in this century and the work of Aston and Dempster. Techniques were subsequently much developed by Bainbridge, Mattauch, Nier, Reynolds and others. As we know it now the instrument may be used in geological dating in a routine way, and its development to this present high level has been responsible chiefly for the immense growth of research in geochronology in the period following 1950.

A mass spectrometer is a device used for determining the isotopic composition of a given element, and for measuring actual amounts of a given isotope. The principles of its operation are straightforward but their practical realization requires elaborate and stable electronics and sound vacuum technology. A simplified diagram of a mass spectrometer is shown in Fig. 3.1. The element being analysed is ionized in the source and these ions are accelerated in falling through a potential difference which might typically be about 2000 V. Narrow defining slits and plates in the ion gun are held at various potentials and serve to collimate the beam of ions. If the element being analysed is strontium, for instance, the beam diverging from the final slit in the source will be composed of Sr^{84}, Sr^{86}, Sr^{87} and Sr^{88} positive ions. This beam passes through a uniform magnetic field aligned perpendicular to the motion of the ions. The charged

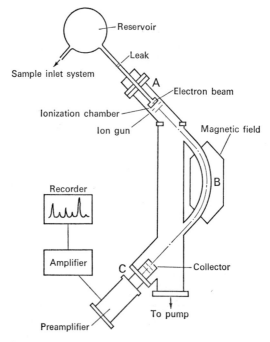

FIG. 3.1. Schematic representation of a typical gas source mass spectrometer.

particles are bent in this field into segments of circular orbits and the net result of the magnetic field is the splitting of the incident beam into four component isotopic beams (one containing only Sr^{84} ions, one only Sr^{86} ions, etc.) each of which, on emergence from the field, comes to a good focus along a line, as shown in Fig. 3.2. A slit located at one of these focal points will permit just one of the ion beams to fall through into a cup, and the charge so collected is neutralized by an electron flow from earth

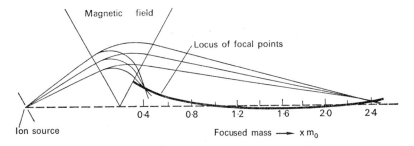

FIG. 3.2. Focusing action of a mass spectrometer. (After Inghram and Hayden, 1954.)

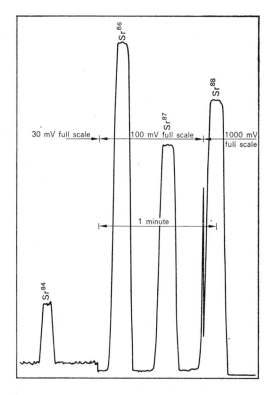

FIG. 3.3. Sr mass spectrum for Moore County achondrite. (After Gast, 1962.)

through a very large resistance (10^{11} ohms, typically). Thus the collected ion beam, which is simply an electric current, produces a voltage across this resistance, is electronically amplified and can be made to produce a signal on a pen recorder. By varying the magnetic field one can swing the orbits of the isotopic ion beams around in space so that first one ion beam, then another, falls through the recording slit. In this way it is possible to record the size of the Sr^{84} beam, then Sr^{86}, Sr^{87} and Sr^{88} and then back again through Sr^{88} to Sr^{84}, repeating this process until the beams have been scanned backwards and forwards across the collector slit perhaps thirty times. Resulting from this will be traces on the pen recorder like those seen in Fig. 3.3. If now we can assume that neither the ionization process in the source nor the detection system discriminated between the various strontium isotopes, then we may deduce that the relative peak heights on the recorder chart are an accurate indication of the relative abundances of the strontium isotopes in the original sample. Higher precision has been obtained in recent years by digital recording.

Two forms of ion source are in common use. For analysing gases, the Bleakney–Nier electron bombardment source is very widely used. In this method the gas is leaked into a chamber across which a stream of electrons is being fired. A certain fraction of the gas molecules is struck and ionized, drawn out of the chamber electrostatically and then accelerated by the large potential drop into the magnetic field.

Many solid elements are ionized by the surface ionization technique. A salt of the solid under analysis is deposited on a tungsten, tantalum or rhenium filament which is mounted in the mass spectrometer source. Emission of ions then takes place from the surface of the heated filament.

In geochronological studies argon and helium are analysed with an electron bombardment source, while potassium, rubidium, strontium, uranium and thorium are customarily studied with the surface ionization method. Lead has been analysed with both types of source—in the gaseous form as lead tetramethyl and in the solid state usually as PbS or lead in boric acid. Much smaller amounts of lead may be analysed by surface ionization, however.

The mass spectrometer assembly is maintained at a pressure of $\gtrsim 10^{-6}$ mm Hg during analysis. This reduces molecular scattering of the ion beams and avoids filament burn-up. Ultra-high vacua, however ($\approx 10^{-9}$mm Hg) are required in the measurement of the argon extracted from very young rocks ($\gtrsim$10 m.y. old) or from potassium-poor older rocks. To achieve these very low pressures, the mass spectrometer has to be evacuated while being baked out at temperatures of 200–400°C for a number of hours. On cooling, the background peaks in the instrument at masses 36, 38 and 40 are zero or negligible and should remain so when the system is isolated from the pumps. Consequently the smaller argon samples may be analysed "statically" in such a system (Reynolds, 1956). A small quantity of the argon is released into the mass spectrometer volume while the system is isolated from the vacuum pumps. The inlet leak is closed and the analysis is carried out on the gas trapped in the system. Much smaller quantities can be analysed in this fashion than in the customary "dynamic" mode wherein the gas is continually flowing through the inlet leak into the source (where only a small fraction is ionized) and all the time is being pumped out of the mass spectrometer tube into the atmosphere.

Mass spectrometers are many and varied. The type we have used for illustration in Fig. 3.1 is a first-order, direction focusing, sector field instrument. The vast majority of geochronological work is carried out with such instruments, usually described as "Nier-type", having sector field angles of 60° or 90°. A certain amount of argon analysis is also carried out with mass spectrometers having 180° deflection of the ion beams (Farrar *et al.*, 1964) and with omegatrons (Grasty and Miller, 1965). For detailed discussions of various types of mass spectrometers reference should be made to Barnard (1952), Mayne (1952), Inghram and Hayden (1954) and Duckworth (1958).

3.3. Isotope Dilution

In an age determination one has to measure the concentrations in rocks and minerals of the *isotopes* of various elements. It is never sufficient merely to estimate the concentration of an element in its entirety (i.e. without reference to its isotopic composition), as one does in the usual chemical analysis.

To resolve such a problem the technique of isotope dilution is commonly used. Consider an element consisting of two isotopes, rubidium for instance. Let us suppose we had some pure Rb^{87}, the amount of which we wished to determine. Using the principle of isotope dilution, we would mix completely this unknown number of Rb^{87} atoms (Rb^{87}) with a very accurately known number of Rb^{85} atoms (Rb_s^{85}). After complete equili-bration of the two sets of isotopes we would then take a portion of this

mixture and analyse it on the solid source mass spectrometer to measure the
Rb^{87}/Rb^{85} ratio $((87/85)m)$ of the resultant rubidium. Clearly, since we have

$$\frac{Rb^{87}}{Rb^{85}_s} = \left(\frac{87}{85}\right)_m,$$

then

$$Rb^{87} = Rb^{85}_s \left(\frac{87}{85}\right)_m. \tag{3.2}$$

The diluting Rb^{85} solution is called "the spike". In actual practice one
never deals with pure mono-isotopic solutions and the equations are
consequently a little more complicated than equation (3.2). Thus natural
rubidium has the two isotopes Rb^{87} and Rb^{85}. Similarly the rubidium spike
has both Rb^{87} and Rb^{85}; however, here Rb^{85} will be heavily enriched
relative to Rb^{87}. If now the number of Rb^{87} atoms extracted from a mineral
is written as Rb^{87}, then the isotope dilution equation (3.2) is modified to

$$Rb^{87} = Rb^{85}_s \frac{\left(\frac{87}{85}\right)_s - \left(\frac{87}{85}\right)_m}{\left(\frac{87}{85}\right)_m \left(\frac{85}{87}\right)_c - 1}. \tag{3.3}$$

The subscripts s, m, c, refer to the spike, the mixture, and common rubidium respectively. The details of the diluting techniques employed with the
various other isotopes are given later.

3.4. K^{40}–Ar^{40} Method

(a) DETERMINATION OF K^{40} CONTENTS

The isotopic composition of naturally occurring potassium has been
found, with few exceptions (Schreiner and Verbeek, 1965), to be sensibly
constant throughout the world, and in all types of geological formation.
Accordingly the K^{40} content of a mineral is most frequently found by a
straightforward chemical analysis for total potassium $(K^{39} + K^{40} + K^{41})$
content followed by multiplication of this quantity by a factor representing
the fraction of K^{40} present in natural potassium. Thus

$$K^{40} \text{ ppm by wt.} = K \text{ ppm by wt.} \times 1 \cdot 22 \times 10^{-4}. \tag{3.4}$$

The most common methods used for potassium determination are flame
photometry (mainly) and isotope dilution. The older classical techniques,

such as that of J. Lawrence Smith, have been widely replaced by flame photometry. Accordingly we shall describe this last method in more detail.

Flame photometry is a simple and straightforward analytical technique and is basically a quantitative version of the old "flame-test". In principle it consists in taking a known weight of the sample, dissolving this in hydrofluoric and either perchloric or sulphuric acids, adding portions of the solution obtained from the residue to a flame and recording the strength of the emission of light of a particular wavelength produced by the potassium

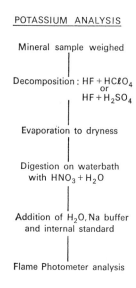

POTASSIUM ANALYSIS

Mineral sample weighed

Decomposition : $HF + HClO_4$
or
$HF + H_2SO_4$

Evaporation to dryness

Digestion on waterbath
with $HNO_3 + H_2O$

Addition of H_2O, Na buffer
and internal standard

Flame Photometer analysis

FIG. 3.4. Outline of flame photometric determination of K.

in the flame. This figure is then compared with those produced by standard solutions. Figure 3.4 shows in outline a typical procedure for potassium measurement with a flame photometer, using lithium as an internal standard.

The major source of error in flame photometry is the depression or enhancement of the potassium emission by extraneous ions. Many people have recorded such effects and numerous conflicting reports exist. A detailed investigation was presented by Cooper (1963) in which he compared the effects of such interference on potassium measurements with several different flame photometric techniques. Figure 3.5 shows the considerable effect of sulphuric acid in the final solution going into the flame. Potassium emission as produced with a Beckman flame spectrophotometer and a Perkin–Elmer single channel flame photometer is strongly suppressed. The measurements with a Perkin–Elmer flame photometer using lithium as an internal standard record a negligible interference effect. The results

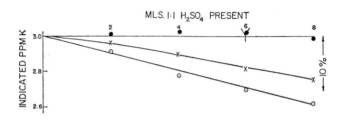

Fig. 3.5. Effect of sulphuric acid on flame photometer reading. × = Beckman; ○ = Perkin–Elmer single channel; ● = Perkin–Elmer internal standard; Y = atomic absorption. (After Cooper, 1963.)

of adding varying amounts of sodium to the solution are seen in Fig. 3.6. In this instance the Beckman results are independent of sodium concentration whereas the Perkin–Elmer data obtained by direct reading and with an internal standard are clearly dependent on sodium concentration. Fortunately the sodium enhancements seem to tend to a constant value, so that by buffering the sample and standard solutions with excess quantities of sodium the effect is minimized. Perchloric acid, iron, magnesium,

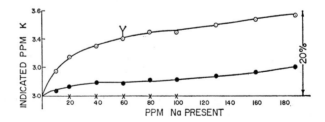

Fig. 3.6. Effect of sodium on flame photometer reading. × = Beckman; ○ = Perkin–Elmer single channel; ● = Perkin–Elmer internal standard; Y = atomic absorption. (After Cooper, 1963.)

aluminium and calcium also interfere with potassium emission and their presence in the final solution may lead to errors. Their effects may be reduced by buffering and by removal of interfering ions. The magnitudes of the enhancements and suppressions seem to be dependent on several parameters, the chief among these being the type of burner used, the composition of the fuel gases and therefore the flame temperature. Several analysts (e.g. Vincent, 1960) recommend a procedure with the E.E.L. (Evans Electroselenium Ltd.) instrument which involves no internal standard and no chemical treatment after the minerals are dissolved. It is supposed that the flame temperature in this instrument is sufficiently low as to avoid interference effects from extraneous ions.

In recent years a number of studies have been published of potassium analysis by isotope dilution. This technique should be at least as accurate as flame photometry and is capable of application to the estimation of lower potassium concentrations. It is, however, a considerably more elaborate method. The chief source of potential errors is the variable mass discrimination common to all solids analysed by surface ionization mass spectrometry.

It should be possible to make K^{40} concentration measurements with an accuracy of $\pm 2\%$. However, at the moment, interlaboratory comparisons indicate that while individual workers can reproduce their own measurements within a spread of $\pm 2\%$, the values obtained may occasionally show up to $\pm 5\%$ difference between laboratories. Some interlaboratory comparisons are shown in Table 3.1.

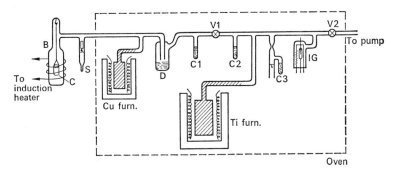

FIG. 3.7. Argon extraction system. C_1, C_2, C_3—charcoal fingers; D—zeolite trap; I.G.—ion gauge; S—spike; V_1, V_2—valves; C—sample crucible. (After Amaral *et al.*, 1966.)

TABLE 3.1. INTERLABORATORY K AND Ar COMPARISONS WITH P-207
MUSCOVITE

(After Lanphere and Dalrymple)

	$K_2O\%$	Ar^{40} $(10^{-9}$ moles/g)
Austral. Natl. Univ.	10·38	1·246
Bundesanstalt für		
Bodenforschung (W. Germ.)	10·38	
Geochron Labs.	10·04	1·231
Geol. Surv. Canada	10·36	1·283
Geol. Surv. Japan	10·37	1·258
Isotopes Inc.	10·23	1·244
Lamont Geol. Obs.	10·38	1·257
Max Planck Inst.	10·39	1·258
Min. Inst. Bern	10·36	
Oxford	10·44	1·264
Socony Mobil	10·42	1·245
Shell Development		1·245
Tohoku Univ.	10·14	1·341
Univ. of Alberta	10·34	1·263
Univ. of Amsterdam	10·24	
Univ. of Arizona	10·43	1·270
Univ. of Calif. Berkeley	10·29	1·265
Univ. of Calif. La Jolla	10·16	1·245
Univ. of Hawaii	9·92	1·254
Univ. of Tokyo	10·40	1·223
Univ. of Toronto		1·273
Penn. State	10·22	
Cambridge Univ.	10·31	
U.S. Geol. Surv.	10·21	1·253
Mean	10·29	1·260
Std. Dev.	0·13 or 1·26% of mean	0·024 or 1·90% of mean

(b) DETERMINATION OF Ar^{40} CONTENTS

Argon is a rare, inert gas occurring in our atmosphere to the extent of
0·93% by volume, the isotopic composition of atmospheric argon being
$Ar^{40} = 99·6\%$, $Ar^{38} = 0·063\%$, $Ar^{36} = 0·337\%$ (Nier, 1950a). The type
of apparatus used for measuring the Ar^{40} content of potassium minerals
is shown in Fig. 3.7. It is an evacuated gas-handling system in which
the pressure before analysis is customarily $\gtrsim 10^{-6}$ mm Hg. The carefully
weighed mineral separate, contained in a metal crucible, is usually baked
out at 100– 200°C for several hours before analysis in order to drive off

moisture and atmospheric gases adhering to the mineral grains. If this is not done, any atmospheric argon trapped on the surfaces of the grains may contribute significant amounts of Ar^{40} during the analysis. While this can be corrected for, it is obvious that the smaller the correction required the better.

The analysis consists simply of (a) melting the mineral sample, with the consequent release of a variety of gases of which H_2O and CO_2 are among the chief constituents and argon is a much less prominent member; (b) purifying the argon by chemical and physical removal of the contaminants; (c) measuring accurately the amount of radiogenic argon released by the molten mineral. In more detail these steps are achieved as follows:

(a) *Fusion of sample.* Radio-frequency induction-heating is the method most commonly used to melt the minerals. In this technique the sample furnace is surrounded by a water-cooled metal coil. In the coil a radio-frequency electrical oscillation is maintained, so that eddy currents are induced in any metal lying within the coil. These currents serve to heat the metal crucible and hence the sample. If the crucible is heated to about 1400°C, there is usually no question that all the radiogenic argon will be expelled from the molten minerals. The sample is held at this high temperature for 15–30 minutes. Meteorites and some minerals such as sanidine and nepheline, however, require a more severe treatment.

Alternatively, the sample-holder may be heated with a wire-wound heater and a flux (e.g. NaOH) is then generally used to melt the minerals. This method has the obvious drawback of requiring careful removal of atmospheric argon from the flux before analysis.

(b) *Purification of the argon.* Easily condensible gases and vapours like CO_2 and H_2O are readily removed from circulation by condensation in the liquid-air cold traps. The oxidizing furnace will convert CO to CO_2 and H_2 to H_2O with their subsequent condensation. Any remaining CO and N_2 and some H_2 are then removed by reaction in the hot titanium furnace. Titanium sponge is a "getter" and will soak up gases (with the exception of inert ones) when hot. (Barium and calcium metal are occasionally substituted for titanium.) After the cooling of the titanium furnace there is essentially argon remaining with perhaps a certain amount of H_2. The latter can be removed by absorbing the argon on activated charcoal, cooled to liquid air temperatures, while the H_2 is pumped away from the system. Recently some workers have used molecular sieves to remove water vapour.

(c) *Measurement of the radiogenic argon.* The argon present in the gas handling system will be a mixture of the radiogenic argon expelled from the minerals plus any contaminating atmospheric argon which has in some way been introduced during the experiment. This latter argon contributes Ar^{40}, Ar^{38} and Ar^{36} in known amounts relative to each other. Consequently its presence is detected by the monitoring of the mass-36 area in the mass spectrometer.

It is most usual to determine the volume of radiogenic Ar^{40} released by the mineral by isotope dilution, using a "spike" consisting of very pure Ar^{38} which may have tiny traces of Ar^{40} and Ar^{36}. During the fusion run a precisely known volume of "spike" is released into the vacuum system where it equilibrates with the radiogenic and atmospheric argon components. This spiked argon is removed from the fusion system after the purification is complete and is examined with a gas-source mass spectrometer. The isotope ratios 40/38 and 36/38 are measured and the volume of radiogenic Ar^{40} released by the mineral is calculated from the expression

$$V_R^{40} = V_S^{38} \left\{ \frac{(40)}{(38)_M} - \frac{(40)}{(38)_S} - \frac{\left[\dfrac{(40)}{(38)_A} - \dfrac{(40)}{(38)_M}\right]\left[\dfrac{(36)}{(38)_S} - \dfrac{(36)}{(38)_M}\right]}{\dfrac{(36)}{(38)_M} - \dfrac{(36)}{(38)_A}} \right\}$$

(3.5)

where V_R^{40} = volume of radiogenic Ar^{40}, V_S^{38} = volume of spike Ar^{38}. Subscripts M, S, A refer to the ratios measured in the final mixture, the spike and atmospheric argon respectively.

The percentage atmospheric argon contamination may be written as

$$\frac{\text{Volume of atmospheric } Ar^{40}}{\text{Atmos. } 40 + \text{Spike } 40 + \text{Radiogenic } 40} \, 100\%$$

$$= \frac{\dfrac{(40)}{(38)_A} \left\{ \dfrac{(36)}{(38)_S} - \dfrac{(36)}{(38)_M} \right\}}{\dfrac{(40)}{(38)_M} \left\{ \dfrac{(36)}{(38)_S} - \dfrac{(36)}{(38)_A} \right\}} \, 100\%$$

(3.6)

If, as is often the situation, $(36/38)_S \ll (36/38)_A$ and $(36/38)_S \ll (36/38)_M$, equation (3.6) becomes

$$\text{Percentage atmospheric Ar}^{40} = \frac{(40)}{(36)_A} \frac{(36)}{(40)_M} \, 100\%. \qquad (3.7)$$

A typical mass spectrum from an argon analysis by isotope dilution is shown in Fig. 3.8 (Farrar, Macintyre, York and Kenyon, 1964).

Since only a few grams of sample are available usually, it is obvious that extremely small volumes of gas are being handled ($10^{-3} - 10^{-8}$ cm^3 n.t.p.). It is therefore customary to make both the fusion system and the mass spectrometer bakeable so that ultra-high vacua may be obtained in both.

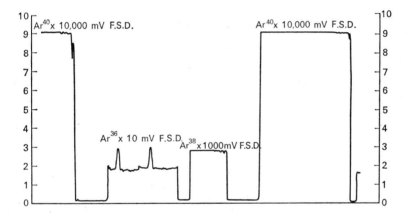

FIG. 3.8. Mass spectrum from argon isotope dilution run on the Toronto MS10.

In this way atmospheric argon contaminations are kept down and the mass spectrometry may be done by the static method mentioned earlier. Assuming that one has a very sensitive and accurate mass spectrometer, it is in fact the atmospheric argon contamination which limits the accuracy of the final calculation of radiogenic argon released from the mineral. The importance of this effect is shown by Fig. 3.9, where it may be seen how, for given errors in the Ar^{40}, Ar^{38} and Ar^{36} peak heights, the percentage error in the radiogenic argon value increases with the atmospheric argon contamination. Once the latter exceeds about 70%, the error in the radiogenic argon determination increases rapidly. If the contamination exceeds 90%, then accurate values for the radiogenic argon volume may only be found if the peak heights are measured with greater accuracies than 0·5%. Furthermore, the Ar^{36} peak is usually several orders of magnitude less than the Ar^{38} and Ar^{40} peaks and is consequently more difficult to measure accurately. The great influence of uncertainty in this peak height is clearly illustrated in Fig. 3.9. For analyses of biotites older than 100 m.y., the atmospheric argon contaminations reported are usually less than 10% and the effect of the contamination is therefore small. For many young basaltic rocks, however, the atmospheric contaminations reported run from 50% to over 90% and the effect described becomes very important.

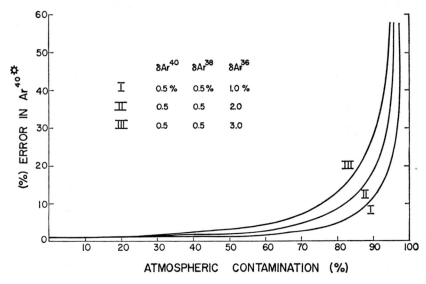

FIG. 3.9. Error in radiogenic Ar determination as a function of atmospheric argon contamination. (After Baksi *et al.*, 1967.)

From equation (3.5) it is obvious that the radiogenic argon volume can not be known more accurately than is the spike volume, V_S^{38}. It is therefore imperative that the latter be known as accurately as possible. In practice this usually turns out to mean an uncertainty of $\pm 1\%$ in V_S^{38}. The spikes themselves are typically about 5×10^{-6} cm³ n.t.p. and are prepared in one of two standard methods, which we call the manifold approach and the reservoir approach respectively. In the manifold way, a large number (anything from 60 to 400) of breakseals of known geometrical volumes are filled simultaneously with spike argon at a single pressure. Thus if the spike content of any one of these breakseals is known, the contents of all of them follow simply from the ratios of geometrical volumes of the breakseals. Accordingly a randomly chosen spike is calibrated by isotope dilution with an accurately known volume of atmospheric argon, and one is then left with a large number of calibrated spikes. To improve the accuracy it is usual to perform calibrations on several spikes to obtain a mean calibration factor. A detailed description of the manifold approach may be found in Wasserburg and Hayden (1955). In the alternative, reservoir approach the spike is contained in a large bulb ($\sim$2 l.) of accurately known volume which is attached to the fusion system. By connecting this bulb to a capillary tube of known volume a fixed fraction of the spike may be extracted for use during a fusion. Since the reservoir pressure is depleted with every run, each spike is smaller than its predecessor. However, if any one aliquot is of a known volume then all the remaining sequence is known, provided an accurate count is kept of the number of spikes doled out. Thus one of the spike aliquots is mixed with a known volume of atmospheric argon in an isotope dilution analysis, which fixes one of the spike volumes and therefore all of them in the numbered sequence. Usually such calibration runs are carried out at intervals to monitor the progress of the system and also to improve the accuracy of the calibration. A detailed description of this method is given by Lanphere and Dalrymple (1966). These two differing approaches to spike preparation are probably about equally widely used at the present time.

Both radiogenic argon and potassium concentrations may be determined with a precision of $\pm 2\%$ in individual laboratories. This would mean a precision of about $\pm 3\%$ in the ratio Ar^{40}/K^{40}. The effect of this uncertainty on the age of a mineral depends upon the absolute value of the age concerned. Thus it would produce uncertainties of about $\pm 3\%$, $\pm 3\%$, $\pm 2 \cdot 6\%$, $\pm 1 \cdot 8\%$ and $\pm 1 \cdot 7\%$ in minerals 10, 100, 500, 1000, 2000 and 2500 m.y. old. Because of the difficulty of eliminating systematic errors it is probably safer to assume that reported Ar^{40}/K^{40} ratios are potentially in error by $\pm 5\%$, which would yield percentage errors of ± 5, ± 5, $\pm 4 \cdot 4$, $\pm 3 \cdot 9$, $\pm 3 \cdot 1$ and $\pm 2 \cdot 8$ at the ages just quoted (see Table 3·1). As we have seen earlier, errors may rise significantly above these levels if atmospheric argon contaminations exceed 70%.

With the methods described, the K–Ar technique has been applied successfully to the dating of rocks and minerals ranging in age from 4·5 b.y. for stone meteorites to 5000 y for terrestrial flows (Dalrymple, 1967; Evernden and Curtis, 1965). While the analytical techniques described here are by far those most widely used, several other methods are in use.

For potassium these include X-ray fluorescence (Hahn-Weinheimer and Ackermann, 1963) and neutron activation (Stoenner and Zähringer, 1958), while argon has also been determined by neutron activation (Armstrong, 1966a).

It has been pointed out by York, Baksi and Aumento (1969) and McDougall and Stipp (1969) that the analysis of K–Ar data from samples with very low radiogenic argon concentrations is often facilitated by plotting the values of Ar^{40}/Ar^{36} versus K^{40}/Ar^{36}. A group of samples of a single age will give a straight line on such a plot from whose slope one can immediately calculate the age. Such a plot is a useful visual aid and has the advantage of being amenable to straightforward statistical analysis (York, 1969). The intercept on the Ar^{40}/Ar^{36} axis may be used as a measure of the atmospheric argon ratio characteristic of the mass spectrometer involved, and the presence of viscous flow when the sample leaks into the mass spectrometer tube may be monitored. The mathematical analysis underlying these plots is analogous with that used in Rb–Sr and U–Pb isochron construction.

3.5. Ar^{39}–Ar^{40} Method

Wänke and Konig (1959) developed a method of K–Ar dating based on neutron irradiation and counting techniques. The fast neutron reaction $K^{39}(n,p)$ produces Ar^{39} while the $Ar^{40}(n,\gamma)$ reaction generates Ar^{41}. From counting the Ar^{39} and Ar^{41} activities produced by neutron bombardment of a potassium-bearing specimen one therefore obtains a measure of the potassium and argon concentrations. The K–Ar age of the irradiated sample may then be calculated. The need for a detailed knowledge of the energy spectrum of the neutron flux and the excitation function of the $K^{39}(n,p)$ reaction may be avoided by subjecting a mineral of known K–Ar age to the same irradiation as the unknown. No correction for atmospheric argon contamination is made in this technique, however, which is a severe limitation. Merrihue (1965) proposed that the Ar^{39} and Ar^{40} should be measured mass spectrometrically so that the Ar^{36} could also be detected and the atmospheric Ar^{40} contamination could be estimated as usual. Merrihue and Turner (1966) established the feasibility of this method. The age of the unknown, T_u, may be calculated from the expression

$$\frac{(Ar^{40}/K^{40})_u}{(Ar^{40}/K^{40})_s} = \frac{(Ar^{40}/Ar^{39})_u}{(Ar^{40}/Ar^{39})_s} = \frac{\exp(\lambda T_u) - 1}{\exp(\lambda T_s) - 1} \tag{3.8}$$

where the Ar^{40} values have been corrected for atmospheric contamination, the subscript s refers to the standard irradiated simultaneously with the unknown, the subscript u refers to the unknown and λ is the decay constant of K^{40}. The Ar^{40} and K^{40} concentrations are thus determined on the same material and the method is applicable to very small samples. Its great appeal, however, lies in its potential for revealing both the thermal histories of minerals and the presence of inherited argon. Turner et al. (1966) carried out stepwise heating of the irradiated Bruderheim chondrite and by examining the form of the release pattern of the ratio Ar^{40}/Ar^{39} as a function of the fraction of the Ar^{39} released found strong evidence that this meteorite was heated 495 ± 30 m.y. ago, at which time it suffered a loss of 90 % of its radiogenic Ar^{40}.

For detailed descriptions of the various factors involved in K–Ar dating reference should be made to McDougall (1966) and Schaeffer and Zähringer (1966).

3.6. Rb^{87}–Sr^{87} Chemistry

The age of a mineral or a piece of rock may be determined by this method from the appropriate version of equation (3.1), which is

$$t = \frac{1}{\lambda} \ln \left\{ 1 + \frac{Sr^{87*}}{Rb^{87}} \right\}, \qquad (3.9)$$

where Sr^{87*} = the number of radiogenic Sr^{87} atoms in some volume of mineral or rock, and Rb^{87} = the number of Rb^{87} atoms in the same volume. When a mineral or rock crystallizes it generally traps a certain amount of strontium in its structure. This is usually called "common strontium", and will often have a composition somewhat similar to that shown on p. 56. Its Sr^{87}/Sr^{86} ratio is usually designated as $(Sr^{87}/Sr^{86})_i$. If we call the present ratio of these isotopes $(Sr^{87}/Sr^{86})_p$, we evidently can write

$$Sr^{87*} = \text{total } Sr^{87} - \text{common } Sr^{87} = Sr^{86} \frac{(Sr^{87})}{(Sr^{86})_p} - Sr^{86} \frac{(Sr^{87})}{(Sr^{86})_i},$$

where Sr^{86} = the number of Sr^{86} atoms in the mineral or rock. We may therefore rewrite equation (3.9) as

$$t = \frac{1}{\lambda} \ln \left\{ 1 + \frac{Sr^{86}}{Rb^{87}} \left[\frac{(Sr^{87})}{(Sr^{86})_p} - \frac{(Sr^{87})}{(Sr^{86})_i} \right] \right\}. \qquad (3.10)$$

Hence, to determine a Rb–Sr age we need to measure the quantities Sr^{86}, Rb^{87} and $(Sr^{87}/Sr^{86})_p$. Furthermore, the initial strontium isotope ratio

$(Sr^{87}/Sr^{86})_i$ must be known. Until 1959 it was almost universal practice to set $(Sr^{87}/Sr^{86})_i = 0 \cdot 71$. Since then, however, it has been customary to carry out experiments which actually define $(Sr^{87}/Sr^{86})_i$, thus eliminating needless guesswork. This new approach is described in Chapter 4.

The three quantities Sr^{86}, Rb^{87} and $(Sr^{87}/Sr^{86})_p$ may all be determined by isotope dilution analysis. Before such analyses are begun, however, potential samples are quickly examined by X-ray fluorescence or optical spectrography so that those with unfavourably low Rb/Sr ratios may be rejected. Strontium and rubidium spikes are added to a small amount of the sample, which is then dissolved in hydrofluoric and perchloric acids in a teflon or platinum beaker. This solution is evaporated to dryness and the residue dissolved in $2 \cdot 5$ N HCl. In solution there will now be rubidium and strontium from both the sample and the spike. Since rubidium and strontium both have isotopes of mass 87 it is necessary to separate these two elements for the mass spectrometry to avoid confusion. The mixture is therefore centrifuged to remove most of the rubidium from solution in the precipitate, while the strontium remains in solution. Further purification is then effected by cation exchange. The strontium-bearing solution is filtered and to it is added some radioactive Sr^{85} which serves as a tracer. This solution is then placed on a cation exchange column which is eluted with $2 \cdot 5$ N HCl. The passage of the strontium down the column is followed by monitoring the Sr^{85} with a counter. After the strontium fraction has been collected and evaporated to dryness, one or two drops of perchloric acid are added and heated to dryness to destroy any organic matter. A few millilitres of 10 N HCl are added and the solution is evaporated to dryness, after which treatment the precipitate is ready to be taken up in a few drops of doubly distilled water and deposited on the centre of a cleaned tantalum filament for the mass spectrometry. The precipitate in the centrifuge tube containing the rubidium is dissolved in HCl and the solution is poured into another cation exchange column, to be eluted with $2 \cdot 5$ N HCl. The rubidium fraction, which may be detected with a simple flame test, is collected for the mass spectrometry.

3.7. Rb–Sr **Mass Spectrometry**

A drop of the rubidium solution is dried on a tantalum filament which is mounted in the mass spectrometer source and the assembly is pumped down to a pressure below 2×10^{-6} mm Hg. The temperature of the filament is raised by increasing the current through it until rubidium ions are detected at the collector. Ten to twenty scans may then be made over the rubidium mass spectrum after the ion beams have spontaneously grown to an easily measureable, steady level. The Rb^{87}/Rb^{85} ratio thus found is substituted in equation (3.3) to give the number of Rb^{87} atoms in the mineral sample. Rubidium analysis is frequently troubled by "memory" effects wherein traces of one sample linger in the source to contaminate the next analysis. Cleaning of the whole source in 6 N HCl followed by rinsings with boiling distilled water removes all trace of it, however.

In strontium analysis, difficulty is sometimes experienced in producing stable ion beams if the filament temperature is raised just as rapidly as in rubidium analysis. Steady ion emission can, however, be obtained by heating the filament for about 8 hours at a temperature just below that of strontium ion emission. Ion beams then tend to appear spontaneously and grow quickly. With coaxing, steady emission with easily measured signals is secured, and the Sr^{87}/Sr^{86} and Sr^{88}/Sr^{86} ratios may be found. A typical scan over the strontium isotope segment of the mass spectrum is seen in Fig. 3.3. Peaks are recorded until from ten to twenty values for each ratio of interest can be measured on the charts.

Clearly it is very time-consuming to do the 8 hours pre-heating in the mass spectrometer. Usually it is carried out in an auxiliary vacuum system in which at least six strontium-loaded filaments may be baked simultaneously overnight. Such pretreated filaments are then removed from the auxiliary unit and analysed in the mass spectrometer one after another. It is then straightforward to do a number of strontium isotope analyses in one day.

The routine production of strong and steady ion beams by the surface ionization method is still to be regarded as involving a fair amount of black magic for some elements. Some workers do not resort to extensive pretreatments, some require electron multipliers for ion detection while others use simple Faraday cup collectors. It is, however, considerably easier to analyse strontium than lead by solid source mass spectrometry.

If a spike strongly enriched in Sr^{86} is used, the desired quantities Sr^{86} and $(Sr^{87}/Sr^{86})_p$ may be calculated from the equations

$$Sr^{86} = Sr_s^{86} \left\{ \frac{\dfrac{(88)}{(86)_M} - \dfrac{(88)}{(86)_s}}{\dfrac{(88)}{(86)_C} - \dfrac{(88)}{(86)_M}} \right\} \qquad (3.11)$$

and

$$\frac{(Sr^{87})}{(Sr^{86})_p} = \frac{(87)}{(86)_M} + \frac{Sr_s^{86}}{Sr^{86}}\left\{\frac{(87)}{(86)_M} - \frac{(87)}{(86)_s}\right\} \tag{3.12}$$

where $Sr_s^{86} =$ the number of Sr^{86} atoms added by the spike, and subscripts $M, s, C =$ final mixture, spike and common, respectively.

A calculation such as this for $(Sr^{87}/Sr^{86})_p$, the present ratio, from the strontium isotope dilution analysis is good enough for minerals such as biotite which have significant radiogenic enrichments of Sr^{87}. However, if $(Sr^{87}/Sr^{86})_p$ is only slightly greater than $(Sr^{87}/Sr^{86})_i$, the initial ratio, then the critical quantity $(Sr^{87}/Sr^{86})_p - (Sr^{87}/Sr^{86})_i$ occurring in the age equation (3.10) can only be accurately known if $(Sr^{87}/Sr^{86})_p$ is measured with great accuracy. Thus suppose $(Sr^{87}/Sr^{86})_i = 0\cdot712$ and $(Sr^{87}/Sr^{86})_p$ is measured to be $0\cdot742$. Then the critical difference between these ratios is $0\cdot030$. However, if the value $0\cdot742$ were in error by $+1\%$, then really $(Sr^{87}/Sr^{86})_p$ should be $0\cdot735$ and $(Sr^{87}/Sr^{86})_p - (Sr^{87}/Sr^{86})_i$ should be $0\cdot023$ and not $0\cdot030$. That is, an error of $+1\%$ in $(Sr^{87}/Sr^{86})_p$ would produce an error of about $+30\%$ in the age obtained. But from equation (3.12) we can see that $(Sr^{87}/Sr^{86})_p$, as determined from an isotope dilution analysis, will be subject to the errors in $(87/86)_M$, $(87/86)_s$, Sr_s^{86} and Sr^{86} and will in practice be usually potentially in error by at least $\pm1\%$. Therefore samples with low radiogenic strontium enrichments require a slightly different treatment from that just described. Rb^{87} and Sr^{86} concentrations must still be found by isotope dilution. The $(Sr^{87}/Sr^{86})_p$ value, however, is best found by measuring this quantity directly with the mass spectrometer, an unspiked portion of the sample being used. Since whole rock samples almost always show low Sr^{87} enrichment, they always require such an approach (or its equivalent by double spiking or spiking with Sr^{84}). A slight modification of the earlier experimental procedure is therefore needed in such cases. Thus the sample is now first dissolved before any spike is added. A known fraction of this solution is then taken and spiked and carried through the procedure described earlier for the determination of the Rb^{87} and Sr^{86} concentrations. The remaining unspiked fraction is taken through the strontium-extraction procedure alone and the resulting strontium-bearing solution is analysed on the mass spectrometer for a direct measurement of the $(Sr^{87}/Sr^{86})_p$ ratio.

While this ratio has now been measured in the most direct possible fashion it will still usually have potential errors of at least $\pm1\%$, for most solid source mass spectrometers have errors of this magnitude due to varying discrimination. However, there is a widely used method for minimizing this effect (Gast, 1962). It is assumed that the ratio Sr^{86}/Sr^{88} in nature is constant and equal to $0 \cdot 1194$. Accordingly, if an analysis of unspiked strontium is being carried out to determine $(Sr^{87}/Sr^{86})_p$, the ratio Sr^{86}/Sr^{88} is measured as well. If this differs from $0 \cdot 1194$, the correction factor is found which brings the measured ratio to this value and then one-half of this factor is used to correct the $(Sr^{87}/Sr^{86})_p$ ratio for discrimination. The so-called "normalized" value of $(Sr^{87}/Sr^{86})_p$ is then

$$\frac{(Sr^{87})^N}{(Sr^{86})_p} = \frac{(Sr^{87})}{(Sr^{86})_M} \quad \frac{1}{2} \left\{ \frac{\dfrac{(86)}{(88)_M} + 0 \cdot 1194}{0 \cdot 1194} \right\} \tag{3.13}$$

where the subscript M refers to quantities measured on the mass spectrometer.

A number of investigations show that this is a useful correction and it is very widely used. At least to a first approximation it seems valid. With such normalization, it appears that $(Sr^{87}/Sr^{86})_p$ values may be measured with single collection systems at least as precisely as one part in seven thousand, and meaningful results may readily be obtained for many whole rock samples.

It should be emphasized that the unspiked strontium mass spectrometer analysis is done for the $(Sr^{87}/Sr^{86})_p$ value so that the Sr^{86}/Sr^{88} ratio in the rock may be measured at the same time and the normalization then applied to the $(Sr^{87}/Sr^{86})_p$. If a Sr^{86} spike is being used, it is always necessary to do the two mass spectrometer runs—spiked and unspiked—if one wishes to have an accurately normalized Sr^{87}/Sr^{86} value. By using a spike enriched in both Sr^{86} and Sr^{84} or simply in Sr^{84}, the need to carry out an unspiked strontium run is removed. From the single such isotope dilution analysis it is possible to calculate accurately both the amount of Sr^{86} present *and* the *normalized* Sr^{87}/Sr^{86} ratio. This is because the Sr^{84} ion beam is now sufficiently enlarged as to be measured with considerable accuracy. Consequently the ratios Sr^{88}/Sr^{84} and Sr^{86}/Sr^{84} may be precisely found and from them an estimate of the fractionation in the isotope dilution run may be made. A thorough discussion of the use of Sr^{84} spiking is given by Long (1966). Fractionation and isotope dilution are discussed by Crouch and Webster (1963) and in considerable detail by Dodson (1963b).

3.8. U–Pb and Th–Pb Methods

The radioactive isotopes U^{238}, U^{235} and Th^{232} decay via long series (see Chapter 4) to Pb^{206}, Pb^{207} and Pb^{208}, respectively. Ages may therefore be calculated from the ratios Pb^{206*}/U^{238}, Pb^{207*}/U^{235} and Pb^{208*}/Th^{232}, where the asterisks indicate a radiogenic component. This method accordingly employs isotope dilution with lead, uranium and thorium spikes. In principle, the weighed sample is dissolved and accurately split into two aliquots. From one of these, the unspiked mineral lead is extracted and its 206/204, 207/204 and 208/204 ratios are measured accurately with a solid source mass spectrometer. To the other aliquot, accurately known amounts of spikes enriched in the Pb^{208}, U^{235} and Th^{230} isotopes are added. Lead, uranium and thorium are then extracted as cleanly as possible for the mass spectrometric determination of the isotope ratios 206/208, 238/235 and 232/230. The number of gram molecules of U^{238}, U^{235} and Th^{232} in the spiked sample solutions are then given by

$$U^{238} = U_s^{235} \left\{ \frac{\dfrac{(238)}{(235)_s} - \dfrac{(238)}{(235)_M}}{\dfrac{1}{137.8}\dfrac{(238)}{(235)_M} - 1} \right\}, \qquad (3.14)$$

$$U^{235} = \frac{U^{238}}{137\cdot8}, \text{ and} \qquad (3.15)$$

$$Th^{232} = Th_s^{230} \left\{ \frac{(232)}{(230)_M} - \frac{(232)}{(230)_s} \right\}, \qquad (3.16)$$

where U_s^{235} and Th_s^{230} are the amounts of these isotopes in the spikes added, measured in gram molecules, and the subscripts s and M refer to spike and final spiked mixture respectively. The total number of gram molecules of mineral lead in the final spiked solution is given by

$$Pb = \frac{Pb_s^{208}}{a^{206}} \left\{ \frac{1 - \dfrac{(206)}{(208)_s}\dfrac{(208)}{(206)_M}}{\dfrac{(208)}{(206)_M} - \dfrac{(208)}{(206)_{\min}}} \right\}, \qquad (3.17)$$

where subscripts s and M have the same meaning as before, "min" refers to the mineral, and a^{206} is the molar fraction of the mineral lead which is Pb^{206}. This latter quantity is defined by

$$a^{206} = \frac{1}{1 + \dfrac{204}{206} + \dfrac{207}{206} + \dfrac{208}{206}},$$

where the ratios are those for the mineral lead. The number of gram molecules of radiogenic Pb^{206} in the final spiked solution is now

$$Pb^{206*} = Pb.a^{206} \left\{ 1 - \frac{\dfrac{(206)}{(204)_c}}{\dfrac{(206)}{(204)_{min}}} \right\} \tag{3.18}$$

where $(206/204)_c$ is the ratio of these isotopes in the "common lead" which is usually present in the minerals being analysed (see Chapter 6), and which will presumably have been trapped at crystallization. The other radiogenic lead components, Pb^{207*} and Pb^{208*}, are given by equations which are derived very simply from the last one by writing the number 207 where there is now 206, and then 208 instead of 206. Thus to make the "common lead" correction we need to know $(206/204)_c$, $(207/204)_c$ and $(208/204)_c$, that is, the isotopic composition of the lead trapped in the mineral at crystallization. An estimate of this is usually made on the assumption that the lead captured at formation has the isotopic composition of lead found in lead minerals such as galena which are of similar age. This is discussed further in Chapter 4. The 207/206 age defined in that chapter is found simply from the quotient $207*/206*$.

In Fig. 3.10 we show an outline of a procedure for the extraction and analysis of uranium, thorium and lead from zircons used by Catanzaro and Kulp (1964). The elegant combination of anion and cation exchange processes is described in great detail by the original authors.

Detailed accounts of the isotope dilution analysis of uranium, thorium and lead without the use of ion exchange methods have been given by Tilton (1951), Tilton *et al.* (1954), Patterson (1951) and Tilton *et al.* (1955). Marshall and Hess (1960) have described a volatilization approach to the analysis of lead in meteorites. The latter are melted in a vacuum and the vaporized lead is condensed in a quartz furnace. Dithizone extractions are then used to purify and concentrate the lead. Various other methods of analysis for these elements are summarized by Hamilton (1965).

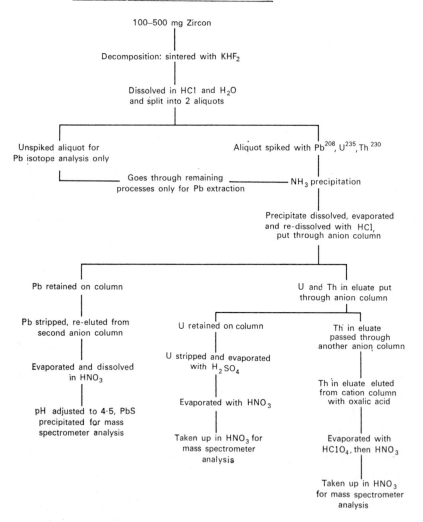

EXTRACTION OF Pb, U AND Th FROM ZIRCON

100–500 mg Zircon

Decomposition: sintered with KHF_2

Dissolved in HCl and H_2O
and split into 2 aliquots

Unspiked aliquot for
Pb isotope analysis only

Aliquot spiked with Pb^{208}, U^{235}, Th^{230}

Goes through remaining
processes only for Pb extraction

NH_3 precipitation

Precipitate dissolved, evaporated
and re-dissolved with HCl,
put through anion column

Pb retained on column

U and Th in eluate put
through anion column

Pb stripped, re-eluted from
second anion column

U retained on column

Th in eluate
passed through
another anion column

Evaporated and dissolved
in HNO_3

U stripped and evaporated
with H_2SO_4

Th in eluate eluted
from cation column
with oxalic acid

pH adjusted to 4·5, PbS
precipitated for mass
spectrometer analysis

Evaporated with HNO_3

Evaporated with
$HClO_4$, then HNO_3

Taken up in HNO_3 for
mass spectrometer
analysis

Taken up in HNO_3
for mass spectrometer
analysis

FIG. 3.10. Outline of U, Th and Pb determinations.

Because of the low lead concentrations found in many minerals of interest this method usually involves strenuous efforts to minimize lead contamination during the analyses. It is usually necessary to invoke such measures as the maintenance of a positive pressure of filtered air in the laboratory, frequent scrubbing of fume-hoods and bench tops, HF digestions in stainless steel or teflon boxes in nitrogen atmospheres, multiple distillations of water and acids, and the routine use of "parafilm" sheet for sealing bottle stoppers. Details of such measures are given by Tilton *et al.* (1955). A zircon might well contain 100 ppm by weight of lead. Thus, if 100 mg are analysed, there will be 10 μg of mineral lead available. If the lead blank in an analysis is 1 μg, then this would represent a 10 % contamination. For meteorite samples which have considerably lower lead concentrations such a blank would be very serious. Significantly lower lead blanks than this have been reported by workers using precautions of the type described above; $0 \cdot 3$–$1 \cdot 3$, $0 \cdot 2$, $0 \cdot 1$ and $0 \cdot 05$ μg of lead blank have been reported by Tilton *et al.* (1957), Doe (1962), Chow and Patterson (1962) and Patterson and Tatsumoto (1964) respectively. Uranium and thorium blanks are usually an order of magnitude less at about $0 \cdot 01$ μg.

The first isotopic analyses of lead and uranium extracted from minerals separated from granites were reported by Tilton *et al.* (1955), who took great strides in the development of the required micro-analytical techniques. The microgram quantities of lead and uranium involved are analysed mass spectrometrically by the surface ionization technique. Lead is usually mounted on a tantalum, tungsten or rhenium filament for ionization either as the sulphide, oxide or oxalate (Cooper and Richards, 1966). Catanzaro and Kulp (1964) and Aldrich *et al.* (1958) brought their final lead nitrate solutions to pH $4 \cdot 5$ before precipitation of PbS by H_2S. Tilton *et al.* (1955) mounted lead on the filament in a solution of borax in nitric acid. Marshall and Hess (1961) have reported good ion emission when lead was mounted on the filament using boric acid. Lead is usually analysed from a single filament source and detected as Pb^+ with the aid of an electron multiplier. The precision of most reported lead isotope ratios seems to be about $0 \cdot 3$–$1 \cdot 0\%$, with the Pb^{204} always being troublesome owing to its low abundance. Uranium and thorium may be analysed as nitrates from single filaments as UO_2^+ and ThO^+ ions, isotope ratios having uncertainties of about 1 %. While the procedure outlined in Fig. 3.10 separates the uranium and thorium so that they may be analysed in different mass spectrometric runs, Aldrich *et al.* (1958) mounted the uranium and thorium together in nitric acid on a tantalum filament. Analysis of the uranium was carried out first since thorium is emitted at a much higher temperature. The uranium was virtually completely evaporated as the filament temperature was increased until thorium emission occurred. In most studies U/Pb and Th/Pb ratios are probably determined with an uncertainty of about 3 %.

The analysis of lead from galenas presents a more straightforward problem, since about 85 % of this mineral is lead and usually large crystals are readily available. Contamination problems are consequently minimal. Furthermore, because of the large quantities of lead involved, many of the reported galena lead analyses have been carried out with gas source mass spectrometers. The most common approach is to convert the lead to liquid lead tetramethyl which provides the vapour for ion production in an electron bombardment source. A complex mass spectrum ensues as the vapour yields lead trimethyl, lead dimethyl, lead monomethyl and lead ions. Hydrogen atoms within these groups may also be lost and further complexity is caused by the existence of two isotopes of carbon, C^{12} and C^{13}. Usually the trimethyl group is chosen for determination of the lead isotope ratios, peaks appearing in the mass range 248–255. This technique, which was pioneered by Aston (1933), requires much larger amounts of lead than does the surface ionization approach. Usually at least 50 mg of lead are taken for conversion to the tetramethyl. However, it does possess the considerable advantage that rapid intercomparison of an unknown lead vapour may be made with a known, in this way reducing significantly the effects of time-varying discrimination between the lead isotopes. With this approach Russell and his co-workers at the University of British Columbia have been able to achieve a precision of approximately $0 \cdot 1 \%$ in lead isotope ratio measurements (Kollar et al., 1960; Russell, 1963). Ulrych and Russell (1964) described a free radical technique for the preparation of lead tetramethyl from samples of about 500 μg of lead and thereby were able to study the isotopic composition of lead from pyrite. The solid source method, however, will operate with two orders of magnitude less lead than this.

Recently Catanzaro (1967) reported obtaining a distinct improvement in the precision of lead isotope ratio analysis with a solid source mass spectrometer. A "triple filament" source was used (Palmer and Aitken, 1953) in which the lead sample was loaded on two rhenium filaments run at a low temperature while the platinum ionizing filament was maintained at 1480°C as read on an optical pyrometer. Ten replicate analyses of a lead solution indicated that the precision for a single analysis, at the 95% confidence level, was $0\cdot054\%$ for Pb^{204}/Pb^{206}, $0\cdot023\%$ for Pb^{207}/Pb^{206} and $0\cdot071\%$ for Pb^{208}/Pb^{206}. The sample was loaded as $Pb(OH)_2$. The only disadvantage seemed to be that larger sample sizes than usual with a solid source seemed to be required ($\sim$ several hundred micrograms).

Because of discrimination effects of various kinds, which affect different mass spectrometers in differing degrees, the agreement in isotope ratios between laboratories is always less promising than the internal results of any one laboratory would predict. For lead isotopes this is illustrated by Fig. 3.11 taken from Russell (1963). The range of variation in reported values of the Ivigtut lead is approximately 3% in 206/204, 207/204 and 208/204 ratios, although most of the laboratories could claim much better internal precision than this. To combat this discrimination several laboratories are developing double-spiking procedures for lead analysis by analogy with the approach adopted in strontium isotope analysis.

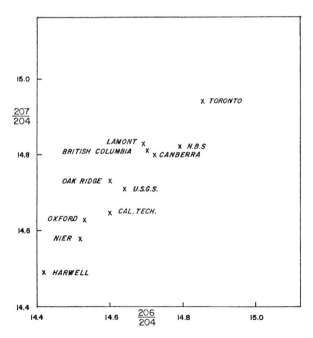

FIG. 3.11. Interlaboratory comparisons of Pb isotope ratio analyses on Ivigtut galena. (After Russell, 1963.)

CHAPTER 4

THE DATING OF UNDISTURBED MINERALS AND ROCKS

4.1. K–Ar Method

Potassium is a common constituent of many rocks (see Table 2.2) and its isotopic composition is usually taken for age studies to be that determined by Nier (1950a): $K^{39} = 93 \cdot 08\%$; $K^{40} = 0 \cdot 0119\%$; $K^{41} = 6 \cdot 91\%$. The rare isotope K^{40} is radioactive and, as we have seen in detail in Chapter 2, has a branching decay to Ca^{40} and Ar^{40}. Because of the widespread abundance of "common" Ca^{40}, however, only the K^{40}–Ar^{40} branch of the decay is used to any extent in age determination. The branching nature of the decay requires that we calculate the K–Ar age of a mineral or rock from a slightly modified version of the fundamental age relation equation (3.1); thus we may write

$$t = \frac{1}{\lambda} \ln \left\{ 1 + \frac{(1 + R)}{R} \frac{Ar^{40}}{K^{40}} \right\}, \tag{4.1}$$

where the "branching ratio" $R = \lambda e/\lambda \beta$, $\lambda = \lambda e + \lambda \beta$, and λe and $\lambda \beta$ are the decay constants for K-capture and β-decay respectively.

The exact isotopic composition of potassium is not easily determined because of the mass discrimination effects which plague solid source mass spectrometry. Averaging the above value from Nier (1950a) with those from Reuterswärd (1952, 1956) and White et al. (1956) gives $K^{40}/K = 0 \cdot 0118\%$, and this value, rather than the more usual $0 \cdot 0119\%$, has been used by a few workers (Folinsbee et al., 1960). Kendall (1960), Létolle (1963) and Kaviladze and Abashidze (1964) found no naturally produced variation in the K^{39}/K^{41} ratio within the limits $\pm 0 \cdot 3\%$, $\pm 0 \cdot 5\%$ and $\pm 0 \cdot 3\%$ respectively. Burnett et al. (1966) found no difference in K^{39}/K^{41} among four terrestrial samples to "within 2 to 3 %". Their values are shown in Fig. 4.1 where they may be compared with those of Nier and Reuterswärd. Burnett et al. were principally searching for any variation in the abundance of the K^{40} isotope. No variation was found, in fact, for terrestrial samples. The only K^{40} variations seen were in the meteorites Norton County, Weekeroo Station and Vaca Muerta and amounted only to about 1 %. In contrast with these indications of constancy of potassium isotopic composition, Schreiner and Verbeek (1965) reported the observation of

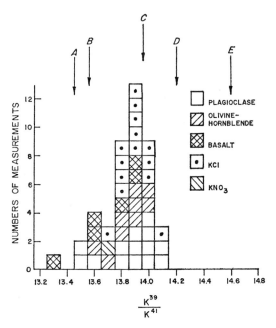

FIG. 4.1. K^{39}/K^{41} values for terrestrial materials. A = Nier (1950a); B = Reuterswärd (1952, 1956); C, D, E = Schreiner and Verbeek (1965). (Modified after Burnett *et al.*, 1966.)

distinct K^{39}/K^{41} variations. Their measurements gave a value of K^{39}/K^{41} equal to $13 \cdot 96 \pm 0 \cdot 04$ for the Cape Granite and $14 \cdot 21 \pm 0 \cdot 03$ for the Malmesbury Sediments, and these they considered to be significantly different. Furthermore, an "unaltered" xenolith of the sediment caught up in the granite had a K^{39}/K^{41} value of $14 \cdot 60 \pm 0 \cdot 05$. These results are illustrated in Fig. 4.1. It is apparent that there is room for many more measurements of potassium isotope ratios in a much wider variety of rocks and minerals. In this connection, Rankama (1954) suggested that minerals capable of cation exchange, such as zeolites, felspathoids, glauconites and clay minerals, might exhibit potassium isotope fractionation. Fortunately, the effect of errors in the assumed isotopic composition of potassium is not as significant as it might seem at first sight. Thus the K^{40}/K ratio influences λe, $\lambda \beta$ and Ar^{40}/K^{40}, so that examination of equation (4.1) shows that variations in K^{40} abundance produce but a small change in age. As pointed out by Smith (1964), diminishing the K^{40}/K ratio by 5% reduces ages of 1, 100, 500, 1000, 2000 and 3000 m.y. by zero, $0 \cdot 1$, $0 \cdot 6$, $1 \cdot 1$, $1 \cdot 9$ and $2 \cdot 4$% respectively.

Although the β-activity of potassium was observed early in the century by Thomson (1905) and Campbell and Wood (1906), it was not until 1937 that von Weiszäcker (1937) proposed that argon was generated by electron capture in K^{40}. He suggested further that old potassium minerals should contain measurable quantities of radiogenic Ar^{40}. After several other investigators had reported unsuccessfully looking for such radiogenic argon, Aldrich and Nier (1948) clearly confirmed von Weiszäcker's suggestion by observing Ar^{40}/Ar^{36} ratios in four minerals which were much greater than the Ar^{40}/Ar^{36} ratio for atmospheric argon. These findings were the touchstone of the tremendous developments in geochronology during the following 20 years. Early applications of the K–Ar age method were made by Smits and Gentner (1950) and Gerling and Pavlova (1951). With the advent of isotope dilution for argon analysis (Inghram et al., 1950) and the development of ultra-high-vacuum mass spectrometry by Reynolds (1956), the K–Ar technique has evolved rapidly until it is now in widespread, routine use in many countries. As a consequence, more geological age determinations have been carried out with this method than with any other.

4.2. Suitability of Minerals for K–Ar Work

The establishment of which minerals are reliable for K–Ar dating is a continuing, vital part of geochronological research. In this section we will discuss the results found for various minerals when these have experienced relatively peaceful post-crystallization histories.

Micas (7–9% K)

Biotite and muscovite are abundant in many acid igneous and metamorphic rocks and most K–Ar age determinations have been made on these minerals. When extracted from rocks which have cooled rapidly and remained subsequently undisturbed, biotite and muscovite have been found to give a reliable K–Ar date of formation. The mica structure evidently is a very favourable one for argon retention. The reasonable concordance of mica K–Ar ages with those obtained by other methods in simple geological settings argues strongly against significant inheritance of Ar^{40} at the moment of crystallization being of widespread importance, and subsequent gradual loss or gain of argon in quiescent conditions is apparently no real problem. Damon and Kulp (1957) found no measureable amount of excess argon in the calcium mica margarite, although the potassium concentration was extremely low and any significant excess would have been simple to observe.

AMPHIBOLES $(0 \cdot 1–1 \cdot 5 \% \text{ K})$

Studies by Hart (1961, 1964), Steiger (1964) and others have clearly demonstrated that hornblende retains argon tenaciously, perhaps more so than the micas (see Chapter 5). In Table 4.1 we show the data of Webb and McDougall (1964) for coexisting biotites and hornblendes separated

TABLE 4.1. K–AR AGES OF COEXISTING PAIRS OF BIOTITE
AND HORNBLENDE

(After Webb and McDougall)

Sample no.	Mineral	K %	Age (m.y.)
781	Biotite	7·08	122
	Hornblende	0·307	123
782	Biotite	6·78	125
	Hornblende	0·296	122
792	Biotite	7·10	128
	Hornblende	0·467	125

from intrusives located west of MacKay in Queensland, Australia. Samples 781 and 782 are from the same Mt. Barker Granodiorite while 792 are from the granitic Urannah Complex less than 10 miles away. The very small age spread within the whole group and within the biotite–hornblende pairs indicates very strongly that the bodies were emplaced essentially simultaneously approximately 125 m.y. ago and have remained subsequently undisturbed. The fact that amphiboles have fairly low potassium contents means that potassium analyses require considerable care.

POTASSIUM FELSPARS (7–12% K)

K-felspars from undisturbed plutonic igneous rocks show unpredictably large losses of Ar^{40}. As much as 60% of the total radiogenic argon may be missing. This effect was first observed by Wetherill *et al.* (1955) and has been confirmed by numerous investigators (e.g. Wasserburg *et al.*, 1956; Goldich *et al.*, 1957). Apart from the fact that K-felspars from plutonic rocks less than 100 m.y. old may often yield reliable ages, there seems to be

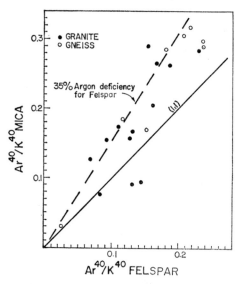

FIG. 4.2. Ar^{40}/K^{40} ratios in micas versus those in associated felspars. (After Goldich *et al.*, 1961.)

no clear correlation between argon loss and age. It is generally considered that the argon escape is connected with the unmixing of the different felspar phases (perthitization) as the crystals adjust to lower-temperature equilibrium conditions (Sardarov, 1957), but no definitive solution has yet been presented. K-felspars from originally deep-seated bodies are therefore currently accepted as being totally unreliable for the dating of such rocks over 100 m.y. in age. This is an acute disappointment since such felspars represent a large component of the felspar group which comprises 60% of the bulk of igneous rocks. The felspar argon anomaly is illustrated in Fig. 4.2 where the data of Goldich *et al.* (1961) show several felspars exhibiting argon deficiencies of about 35% relative to index micas. However, the volcanic, high temperature K-felspar, sanidine, has been shown by Baadsgaard *et al.* (1961) to retain argon as well as comparable biotite. Their data for sanidine and biotite from the upper Cretaceous, Bearpaw Bentonite are given in Table 4.2. Measurements on

TABLE 4.2. COMPARISON OF Ar^{40}/K^{40} RATIO OF BIOTITES
AND SANIDINES FROM BEARPAW BENTONITE
(After Baadsgaard, Lipson and Folinsbee)

Sample	K^{40} (ppm)	Ar^{40}/K^{40}
Biotite 60–100 mesh	6·03	0·00408
Biotite 60–100 mesh	6·03	0·00464
Biotite 60–100 mesh	7·05	0·00440
Biotite 60–100 mesh	7·02	0·00440
Biotite 100–170 mesh	6·67	0·00439
Biotite 170–270 mesh	6·28	0·00450
Sanidine 100–140 mesh	5·48	0·00457
Sanidine 100–170 mesh	10·56	0·00462
Sanidine 100–270 mesh	9·02	0·00461

the diffusion coefficients of argon in sanidines by these same investigators are shown in Fig. 7.2. Sanidines less than 40 m.y. old gave satisfactory ages in a study by Evernden *et al.* (1964).

PLAGIOCLASES (0·01–2% K)

McDougall (1963) obtained reasonable K–Ar ages for plagioclases from dolerites approximately 160–180 m.y. old. Evernden and James (1964) suggested that volcanic plagioclases with potassium concentrations less than 0·9% would retain argon "quantitatively" under near-surface conditions for "hundreds of millions of years." Volcanic felspars of intermediate potassium concentrations, with unmixing tendencies, showed argon deficiencies. In a study of much older plagioclases from British Guiana dolerites, McDougall *et al.* (1963) observed a significant spread in age which they interpreted to mean loss of radiogenic argon at "low temperatures, probably below 200°C". These dolerites were extremely fresh and evinced no trace of having been metamorphosed. There is clearly room for much further research on older volcanic plagioclases and virtually nothing is known of the argon retentivity of plutonic plagioclases.

PYROXENES (0–1 % K)

Because of their low potassium contents pyroxenes have only recently begun to receive detailed attention. It is already apparent, however, that their use requires great care. The evidence to hand indicates that pyroxenes formed at great depth are distinctly susceptible to yielding anomalously high K–Ar ages (Gerling et al., 1962; Hart and Dodd, 1962; McDougall and Green, 1964; Lovering and Richards, 1964). While this is not always the case (Hart, 1961), there is no doubt that pyroxene dates from deep-seated masses must be viewed with reservation until more data clarify the situation. On the other hand, pyroxenes which crystallized near the surface seem to be far more reliable (McDougall, 1963). This is fortunate, in that pyroxene is a common constituent of basaltic lavas which occur widely on all continents.

GLAUCONITES (4–7 % K)

More geochronological work has been performed with this mineral than with any other of sedimentary origin, yet its status is still obscure. Since the early work of Lipson (1956) and Wasserburg et al. (1956), evidence has accumulated which demonstrates that in many instances glauconites give ages which are too low by 10–20 % and, as a consequence, the conclusions drawn from glauconite ages are usually couched in cautious terms (Evernden et al., 1961; Kazakov and Polevaya, 1958; Hurley et al., 1960; Dodson et al., 1964). It appears that glauconite is susceptible both to argon loss and potassium gain, each of which would cause anomalously low ages (Hurley et al., 1960). There is no simple age effect apparent: glauconites younger than 100 m.y. tend to be more reliable, while Webb et al. (1963) found apparently reasonable ages in the 1500 m.y. region; but early palaeozoic glauconites frequently give K–Ar ages which are too low. On balance they must still be considered generally unreliable, but may be reasonably safely considered to give *minimum* dates for the sediments. Hurley (1966) has reviewed the problems involved in the K–Ar dating of sediments.

NEPHELINES (3–7 % K)

The suitability of these minerals for K–Ar work has recently been investigated by Macintyre, York and Gittins (1966), who found that with the samples studied, the nepheline K–Ar age always equalled or was greater than the index mica or hornblende age, as may be seen in Table 4.3. Since

TABLE 4.3. COMPARISON OF K–Ar AGES OF COEXISTING NEPHELINES, BIOTITES AND AMPHIBOLE

(Modified from Macintyre et al., 1966)

Locality	Biotite age (m.y.)	Nepheline age (m.y.)	Amphibole age (m.y.)
Princess Quarry, Ontario	900	904	
Goulding-Keene Quarry, Ontario	855, 836 (C) 865 (F)	879, 903	
Blue Mountain, Ontario	900*	1032, 1053	
Bigwood twp., Ontario	888, 877	1169, 1171	1003, 1007

* Determination by Geol. Surv. Canada.
C = Coarse grain.
F = Fine grain.

the nepheline ages did not exceed Rb–Sr whole rock or U–Pb zircon ages from the same geological province, it was suggested that nepheline may be distinctly more retentive of its argon than are micas and amphiboles. The possibility that nephelines may give anomalously high K–Ar ages can only be ruled out by further analytical work.

OTHER MINERALS

Several other minerals have received a small amount of attention, with results which have been discouraging or inconclusive. These include sylvite (Gentner et al., 1953, 1954; Inghram et al., 1950; Polevaya et al., 1958); calcite (Lippolt and Gentner, 1963); illite (Evernden et al., 1961; Bailey et al., 1962; Hower et al., 1963). Minerals such as beryl which are characterized by anomalously high ages are discussed in Section 4.4.

4.3. Suitability of Whole Rocks for K–Ar Analysis

As is described in Chapter 5, in Rb–Sr dating it is vital to carry out analyses on whole-rock samples. Thus when radiogenic strontium leaks out of one mineral it usually finds its way into another, and if sufficiently large volumes are considered, we may still find a good approximation to a closed system. This is not at all the case with the K–Ar method, unfortunately. When argon escapes from one mineral it apparently does not usually enter one nearby but rather migrates quickly from the area. Dating a whole-rock sample would give an average age which would be a composite of the individual mineral ages. If some of the constituent minerals are poor argon retainers, the whole-rock age will be too low to an extent which depends on the relative abundances of the various minerals and on their relative potassium concentrations. Whole-rock K–Ar dating of a granite would therefore be unsatisfactory because of the presence of significant quantities of potassium felspar which, on account of its almost inevitable argon loss, would reduce the apparent age of emplacement. Wherever possible, therefore, K–Ar dating is carried out on mineral separates and concordancy among different mineral types strongly indicates reliability. However, many volcanic rocks are so fine-grained that mineral separation is prohibitively difficult and in such cases recourse has increasingly been made to whole-rock K–Ar dating. By far the most prominent rock-type studied in this manner is basalt.

WHOLE-ROCK VOLCANICS

Under this heading we include sills, dykes and flows. It is also convenient to restrict our considerations initially to such rocks which are less than about 200 m.y. old. Since the early studies of the Palisades Sill by Erickson and Kulp (1961), evidence has been accruing that K–Ar dating of fine-grained basaltic material in this age range may be satisfactory, provided that rigid control is exercised over sample freshness. Tremendous impetus has been given to the dating of such materials by the desire to resolve the question of the reversals of the earth's magnetic field, a topic discussed in some detail in Chapter 10. Consistent and apparently reliable whole-rock K–Ar dates have been obtained in the 0–10 m.y. age range by Dalrymple and others in numerous studies (Cox et al., 1963a, b; Dalrymple, 1964), by McDougall and Tarling (1963) and by Evernden et al. (1964) and Evernden and James (1964). Undoubtedly the short time available for diffusive losses and for post-crystallization periods of metamorphism in the last 10 m.y., contribute to this. Amaral et al. (1966) found satisfactory ages at approximately 120 m.y. for whole-rock samples of basalt and diabase from the Parana basin of southern Brazil. Measurements by McDougall and Ruegg (1966) on separated felspars and pyroxenes from some of the same localities agreed within experimental error with the data of Amaral et al. Whole-rock diabase dates in the range 150–190 m.y. were found by McDougall (1961, 1963) for diabases from south-east Africa, Antarctica and Tasmania.

It is evident then that this type of whole-rock K–Ar dating has considerable potential for rocks less than about 200 m.y. in age. For basic volcanics older than this it becomes increasingly difficult to find fresh material and correspondingly more difficult to obtain reliable ages on whole-rock samples. Several studies of palaeozoic materials have been reported by Miller and co-workers, their work being summarized by Miller and Fitch (1964). K–Ar whole-rock studies on Precambrian dykes have been reported by Fahrig and Wanless (1963) and Leech (1966). All of these studies clearly define the difficulties and indicate the need for further detailed studies of the reliability of K–Ar dates on whole-rock samples greater than 200 m.y. in age.

Regardless of age, however, it is widely recognized that rigid sample control is required. Careful thin-section analysis is important so that samples showing the minimum trace of alteration may be selected. Discussions of potentially important criteria for whole-rock sample selection may be found in papers by Evernden et al. (1964), Miller and Fitch (1964), Amaral et al. (1966) and Baksi et al. (1967). Amaral et al. (1966) noticed that the whole-rock K–Ar dating of diabases ($\sim$ 120 m.y. old) is "occasionally susceptible to discrepancies of up to 8 % even when fairly rigid thin-section criteria have been exercised in the selection of the samples for dating". In addition these investigators found evidence to support the observation of Erickson and Kulp (1961) that finer-grained samples of basaltic rocks tended to be more argon retentive than coarser-grained equivalents. The observation by Fahrig and Wanless (1963) and Leech (1966), that the most reliable ages for Precambrian diabase dykes are obtained on the chilled margins, is consistent with this effect. Glass-bearing samples are suspect if there is evidence of devitrification of the glass (Evernden and James, 1964), although this is not always the case (Baksi et al., 1967). Fresh, undevitrified glass may, however, retain argon well for long periods. Philpotts and Miller (1963) obtained a reasonable age of 975 $\pm$ 46 m.y. for undevitrified glassy pseudo-tachylite from the Grenville province north of Three Rivers, Quebec, Canada. Their results are shown in Table 4.4. The variation of Ar^{40} and K concentrations within a single basalt lava flow and the precision of analysis that may be obtained is illustrated in Table 4.5 taken from Dalrymple and Hirooka (1965).

COLLEGE OF THE SEQUOIAS
LIBRARY

TABLE 4.4. ARGON RETENTIVITY OF GLASS
(After Philpotts and Miller)

Sample	$K_2O\%$	Age (m.y.)
Glassy pseudo-tachylite	3·39	975 ± 46
Entirely crystalline pseudo-tachylite	2·54	904 ± 43
Partly glass pseudo-tachylite	2·81	822 ± 40

TABLE 4.5. VARIATION OF K, Ar AND AGE WITHIN A BASALT FLOW
(After Dalrymple and Hirooka)

Sample	$K_2O\%$	$Ar^{40}*$ $(10^{-11}$ mole/g)	Age (m.y.)
509–62–2	2·18	1·066	3·31
509–62–3	2·63	1·313	3·38
509–62–4	2·74	1·307	3·24
509–62–5			
no. 2	2·58	1·291	3·39
3V121–0	2·39	1·163	3·30
3V124–0	2·46	1·162	3·20
3V128–0	2·44	1·231	3·41
	Total spread, 0·56 = 22·5%	Total spread, 0·247 = 20·3%	Total spread, 0·21 = 6·33%

While the K^{40} and Ar^{40} concentrations vary within the flow by up to 20%, the age spreads by only about 6%, indicating a reliable mean age.

SLATES

The only other rock type examined to any extent as a whole-rock specimen by the K–Ar method is slate. After a few sporadic dates (e.g. Goldich *et al.*, 1957; Hurley *et al.*, 1959) had appeared in the literature, the first detailed geochronological investigation of slate was made by Dodson (1963a). Slaty cleavage is acquired when an argillaceous sediment responds to folding by recrystallization in such a manner that the cleavages of lamellar minerals are parallel. Most of the potassium will be concentrated in fine-grained muscovite. Since the detrital minerals in the original sediment in many cases will contain significant amounts of radiogenic Ar^{40} (Krylov, 1961; Bailey *et al.*, 1962; Hower *et al.*, 1963), it is imperative that the folding expel all argon which predates the cleavage if the slate is to yield the date of folding. Whether this in fact has happened in any particular case is not easy to say. Dodson (1963a) considered that the following criteria were useful in selecting slates for K–Ar whole-rock dating: (a) the slate should be well cleaved in hand specimen; (b) there should be good optical alignment of the minerals in thin section; (c) a sharp, well-defined 10 Å peak should be visible on an X-ray diffraction pattern. Harper (1964), in a study of the British Caledonides, concluded that whole-rock K–Ar dating of slates is a powerful tool in the unravelling of complex cooling histories. Despite the fine grain of the slates, argon was clearly retained by the micaceous minerals over a period of 400 m.y.

4.4. Minerals giving High K–Ar Ages

The first observation of an excessive amount of a radiogenic, gaseous isotope in a mineral was recorded in 1908 by R. J. Strutt (later Lord Rayleigh). Beryls were demonstrated to contain quantities of helium far in excess of those which would have been generated by the radioactive decay of the uranium and thorium in the minerals. A quarter of a century later, Rayleigh (1933) returned to the problem and claimed to have shown an "age effect" wherein the older the beryl the greater was the volume of the excess helium. This conclusion was disputed by Khlopin and Abidov (1941). The problem was given a new facet by Aldrich and Nier (1948), who found that beryls also contain anomalously large concentrations of Ar^{40}. Damon and Kulp (1958), in a fundamental study, confirmed this observation and also recorded excesses of both Ar^{40} and He^4 in five cordierites and two tourmalines. Furthermore, their data not only supported Rayleigh's suggestion of an "age effect" for He^4 but also indicated the same effect for excess Ar^{40}. More recently, anomalously large amounts of Ar^{40} have been found in pyroxene (Hart and Dodd, 1962; McDougall and Green, 1964; Lovering and Richards, 1964; Baadsgaard *et al.*, 1964), quartz (Rama *et al.*, 1965), fluorite (Lippolt and Gentner, 1963), the felspars plagioclase and albite (Livingston *et al.*, 1967), and the felspathoids cancrinite and sodalite (York *et al.*, 1969). The ranges of excess Ar^{40} found in these minerals are shown in Table 4.6.

TABLE 4.6. MINERALS EXHIBITING EXCESS
ARGON CONTENTS

(After York *et al.*)

Minerals	Excess Ar^{40} conc. 10^{-5} cm^3 n.t.p./g
Beryl	3200–4·5
Cordierite	80 –14
Cancrinite	26–24
Tourmaline	12–1·1
Sodalite	1·5–0·59
Quartz	1·2–0·004
Pyroxene	1·1–0·0
Plagioclase	0·27–0·03
Albite	0·07–0·03
Fluorite	0·02–0·0

Damon and Kulp (1958) emphasized the similarities in the structures of beryl, cordierite and tourmaline, each being based on a six-membered tetrahedron ring. In beryl and cordierite these hexagonal rings are stacked in such a manner as to yield a series of channels parallel to the Z-axis. The channels in both minerals harbour varying amounts of alkali ions and water molecules and Damon and Kulp considered that the vast majority of the excess argon was also held there. Strong support for this proposal was provided by the work of Schreyer *et al.* (1960) and Smith and Schreyer (1962) who found, from X-ray powder diffraction studies, that argon atoms introduced into synthetic cordierites at high pressures and temperatures (10,000 bars, 900°C for 5 hours) were located at the centres of the large spherical cavities which are the basis of the cordierite channel structure. These cavities have an average radius of 2·2 Å, are linked by apertures of radius 1·4 Å, and are easily able to accommodate the argon atom whose radius is 1·9 Å. York *et al.* (1969) pointed out that the framework silicates sodalite and cancrinite are strikingly similar in having channels in their structures (Deer *et al.*, 1964), and suggested that the excess Ar^{40} contents of cancrinite and sodalite are contained within their channels, as is the case with beryl and cordierite. It was also proposed that the occurrence of excess argon in all these beryl-like minerals has the same explanation, although the explanation itself remains uncertain. Two obvious possibilities exist: either the argon was incorporated at the time of crystallization of these minerals, or the argon has diffused into them during their post-crystallization history. The first possibility was favoured by Damon and Kulp in the case of beryl, cordierite and tourmaline, whereas Schreyer, Yoder and Aldrich preferred the second. In Chapter 7 it is considered that many minerals do not retain argon during their early high-temperature existence, argon retention only commencing effectively when the temperature has dropped below some threshold "blocking temperature" for a particular mineral. It may well be that this argon lost by the majority of minerals is soaked up in some way by the beryl-like minerals, the top five in Table 4.6. On this model it is possible to explain the Rayleigh age effect in the following way. Old orogenic belts represent deep levels of erosion and therefore, on average, contain minerals which were at great depths and high temperature for a long time. Consequently most of their minerals may have been losing all their argon for periods up to several hundred million years after crystallization. Beryl-like minerals in such an environment would therefore be expected to acquire large amounts of Ar^{40}. Young orogenic belts, on the other hand, show only shallow depths of erosion. Their high-potassium minerals will have cooled quickly to temperatures at which argon is retained and beryls from such belts would therefore tend to capture only small volumes of radiogenic Ar^{40}.

The possibility also exists of the anomalously high K–Ar ages being due to loss of potassium from the minerals. As may be seen from Fig. 7.1, if potassium diffused from an assumed spherical crystal with the diffusion parameter $D/a^2 > 6 \times 10^{-18} \text{ sec}^{-1}$, then apparent ages greater than that of the earth would be produced. Furthermore, Barrer and Falconer (1956) have shown that synthetic basic sodalite and cancrinite are similar to zeolites in their capacity for ion exchange, so that loss of potassium by some mechanism is possible. However, Damon and Kulp's (1958) Beartooth beryl would have to contain over 100% potassium of normal isotopic composition for the age anomaly to be removed. As such a potassium content is impossible, then that particular beryl undoubtedly owed its anomalous age to an excess of argon and not a deficiency of potassium. Accordingly it seems unnecessary to invoke potassium loss in general.

The five remaining minerals in Table 4.6, quartz, pyroxene, plagioclase, albite and fluorite, are characterized by lower volumes of excess Ar^{40} than are the beryl-like minerals. Much of their excess Ar^{40} may be trapped in fluid inclusions (Rama et al., 1965) or in defects in the crystal structures (Hart and Dodd, 1962).

A final point which may be made is that all the minerals in Table 4.6 usually have low potassium concentrations, i.e. much less than 1%.

The study of the characteristics of the occurrence of excess argon in minerals is of considerable importance in K–Ar dating since in principle any mineral is capable of exhibiting such excesses given the appropriate environment.

4.5. Rb–Sr Method

Rubidium has two isotopes and the isotopic composition usually taken in age studies (Nier, 1950b) is $Rb^{85} = 72·15\%$; $Rb^{87} = 27·85\%$; corresponding to a ratio of $Rb^{85}/Rb^{87} = 2·59$. Earlier values obtained are given in Rankama (1954). The most comprehensive study of the isotopic composition of rubidium is that of Shields *et al.* (1963), who examined rubidium extracted from twenty-seven silicates ranging in age from 20 to 2600 m.y. and found an average value of "$Rb^{85}/Rb^{87} = 2·5995 ± 0·0015$". No sample was found to differ significantly from this mean. It would seem that the uncertainty in this isotope ratio is therefore not great enough to be important in age calculations. One of the two isotopes, Rb^{87}, is radioactive and decays to Sr^{87} with the emission of a low-energy β-particle. The uncertainties surrounding the value of the Rb^{87} decay constant have been discussed earlier in Chapter 2. Suffice it to say here that one of the two values $\lambda = 1·47 \times 10^{-11}$ y^{-1}, or $1·39 \times 10^{-11}$ y^{-1}, is used. It is obvious that the uncertainty in our knowledge of this constant is a disconcerting feature of Rb–Sr dating, since a given percentage error in the decay constant causes an error of the same magnitude, but opposite sign, in the calculated age.

When a mineral crystallizes, it will usually incorporate both rubidium and strontium ions, although the ratio of Rb/Sr and the absolute amount of these ions will vary widely, depending on the mineral type. It is customary to refer to this initial strontium as "common strontium". The isotopic composition will in the vast majority of cases not vary widely from $Sr^{88} = 82·56\%$; $Sr^{87} = 7·02\%$; $Sr^{86} = 9·86\%$; $Sr^{84} = 0·56\%$. The value of the ratio Sr^{87}/Sr^{86} in this common strontium is thus generally not far from $0·712$, and indeed, up to 1959 was almost universally assumed to have essentially this value for age calculations. To allow for the presence of the Sr^{87} trapped at crystallization we calculate the Rb–Sr age of a mineral from equation (3.10) derived in Section 3.6, namely

$$t = \frac{1}{\lambda} \ln \left\{ 1 + \frac{Sr^{86}}{Rb^{87}} \left[\frac{(Sr^{87})}{(Sr^{86})_p} - \frac{(Sr^{87})}{(Sr^{86})_i} \right] \right\}, \qquad (4.2)$$

where Sr^{86} and Rb^{87} are the atomic concentrations of these isotopes in the mineral, $(Sr^{87}/Sr^{86})_p$ is the present-day value of this isotopic ratio and $(Sr^{87}/Sr^{86})_i$ is the Sr^{87}/Sr^{86} ratio in the common strontium present in the mineral since its genesis. The quantity

$$Sr^{86} \left[\frac{(Sr^{87})}{(Sr^{86})_p} - \frac{(Sr^{87})}{(Sr^{86})_i} \right]$$

represents the radiogenic Sr^{87} which has been generated by the Rb^{87}. Until 1959, ages were obtained by *measuring* Sr^{86}, Rb^{87} and $(Sr^{87}/Sr^{86})_p$, *assuming* $(Sr^{87}/Sr^{86})_i = 0 \cdot 712$ (or thereabouts) and substituting these quantities in equation (4.2). However, if the Sr^{87}/Sr^{86} ratio has not increased much during the mineral's lifetime, the quantity $(Sr^{87}/Sr^{86})_p - (Sr^{87}/Sr^{86})_i$ will be very small and it is extremely important to know the value of $(Sr^{87}/Sr^{86})_i$ with great accuracy. It is obviously imperative to remove this assumption of a common $(Sr^{87}/Sr^{86})_i$ value and this is, in fact, achieved in the following fashion. Equation (4.2) may be rewritten as

$$\frac{(Sr^{87})}{(Sr^{86})_p} - \frac{(Sr^{87})}{(Sr^{86})_i} = \frac{Rb^{87}}{Sr^{86}} (e^{\lambda t} - 1), \qquad (4.3)$$

which, for constant $(Sr^{87}/Sr^{86})_i$ and t, shows that there is a linear relationship between $(Sr^{87}/Sr^{86})_p$ and Rb^{87}/Sr^{86} (Nicolaysen, 1961). Thus, (a) if micas, felspars and *whole-rock* samples are taken from various parts of the same rock body which was formed essentially instantaneously at time t m.y. ago; (b) if the $(Sr^{87}/Sr^{86})_i$ value was constant throughout the original rock mass; (c) if the individual minerals have all remained closed systems as far as rubidium and strontium movement is concerned; (d) if the quantities Rb^{87}/Sr^{86} and $(Sr^{87}/Sr^{86})_p$ are measured for the various minerals and whole rocks; then the corresponding points plotted on a graph of $(Sr^{87}/Sr^{86})_p$ vs. Rb^{87}/Sr^{86} will lie on a straight line. The time of formation of the rock will be given by

$$t = \frac{1}{\lambda} \ln \{ 1 + slope \}, \qquad (4.4)$$

and the initial strontium isotope ratio, $(Sr^{87}/Sr^{86})_i$, is now *calculated* from the relation

$$\frac{(Sr^{87})}{(Sr^{86})_i} = \text{intercept on } \frac{(Sr^{87})}{(Sr^{86})_p} \text{ axis.}$$

This result is illustrated in Fig. 4.3. In this manner, the age of a rock is found without resort to any assumption regarding the numerical value of $(Sr^{87}/Sr^{86})_i$. The latter parameter is, in fact, now measured and we are

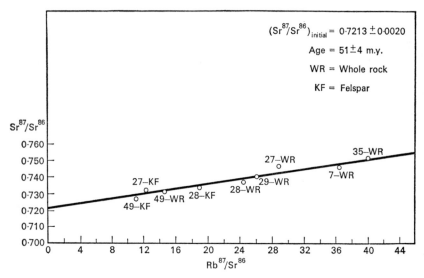

FIG. 4.3. Whole-rock and mineral Rb–Sr isochron. (After Moorbath and Bell, 1965.)

thus able to compare $(Sr^{87}/Sr^{86})_i$ values for various rock types and ages and we have, furthermore, a new tool for the study of petrogenesis. This mode of approach to Rb–Sr dating was originated by Schreiner (1958), Compston and Jeffery (1959), and Nicolaysen (1961) and has been adopted by all the workers in the field.

In requirement (a) of the last paragraph we emphasized "whole rock". The reason for this is simply that when Sr^{87} leaks from one mineral it usually ends up in another. Thus, if sufficiently large volumes are taken, whole-rock systems may often remain closed to overall movement of isotopes, despite the occurrence of ionic migration between individual minerals.

While Rb–Sr dating in this way, involving the analysis of several whole-rock samples and various minerals, is much more laborious than before, the additional information and confidence gained more than compensate for this. It is now generally recognized that it is insufficient to "date" a rock merely by performing a Rb–Sr analysis on an extracted biotite and assuming some intuitive $(Sr^{87}/Sr^{86})_i$ value. Since we are discussing ideally undisturbed systems in this chapter, the transcendent importance of the whole-rock system in Rb–Sr dating may not be unduly apparent. In the next chapter the situation will be considerably different.

In order to get accurate estimates of both the age of a rock and its $(Sr^{87}/Sr^{86})_i$ value, it is obviously important to have a reasonable distribution of points on a diagram such as Fig. 4.3, which is called an "isochron" plot. Several points with low $(Sr^{87}/Sr^{86})_p$ values help to fix $(Sr^{87}/Sr^{86})_i$ within narrow limits, while samples having considerable Sr^{87} enrichments facilitate the precise assignment of an age. Thus before any mass spectrometric work is begun, numerous samples of minerals and whole rocks are analysed rapidly by optical spectrography or X-ray fluorescence for Rb/Sr to enable the best isochron spread of points to be found. Since in many instances samples of low Sr^{87} enrichment predominate, the reconnaissance usually assumes the nature of a search for those few samples with reasonably high $(Sr^{87}/Sr^{86})_p$ values.

A number of minerals have been studied for their suitability, the chief criterion being the Rb/Sr value. Those minerals which are rich in strontium relative to rubidium serve mainly to fix the $(Sr^{87}/Sr^{86})_i$ value and are not otherwise useful in determining the age of a rock.

4.6. Suitability of Minerals for Rb–Sr Work

If we set $e^{\lambda t} - 1 = \lambda t$ in equation (4.3) we can write the approximation

$$\frac{(Sr^{87})}{(Sr^{86})_p} - \frac{(Sr^{87})}{(Sr^{86})_i} \simeq 4 \times 10^{-5} \frac{Rb}{Sr} t, \tag{4.5}$$

where Rb/Sr is the weight ratio and t is in millions of years. If we consider that to have a good spread of points on an isochron plot we would like to have some samples with $(Sr^{87}/Sr^{86})_p$ at least equal to $0\cdot8$, then we desire $(Sr^{87}/Sr^{86})_p - (Sr^{87}/Sr^{86})_i \gtrsim 0\cdot1$ and so from equation (4.5) we require

$$\frac{Rb}{Sr} t \gtrsim 2\cdot5 \times 10^3. \tag{4.6}$$

Thus if we are examining a mineral 100 m.y. old we call for a ratio of Rb/Sr $\gtrsim$ 25. Such a value is found in biotites, but less frequently in muscovites, felspars and whole rocks.

MICAS

Under ideally quiescent post-crystallization conditions radiogenic strontium is apparently well retained in both biotites and muscovites and these minerals are the most commonly analysed in Rb–Sr work. Lepidolite usually has an exceptionally favourable Rb/Sr ratio but is too rare to be really useful. Under metamorphic conditions the Rb–Sr age of a biotite is apparently more readily affected than that of muscovite (Aldrich *et al.*, 1965). In Table 4.7 the data of Zartman (1964) are shown to illustrate the type of concordance which may be found in simple cases.

TABLE 4.7. Rb–Sr and K–Ar Ages from Lone Grove Pluton Area

(After Zartman)

Rb–Sr ages (m.y.)

	Average age	Spread
Microcline	1025	980–1065
Muscovite	1030	960–1075
Biotite	1010	975–1025

K–Ar ages (m.y.)

Muscovite	1050	1020–1080
Biotite	1045	1025–1075
Hornblende	1045	980–1080

FELSPARS

The study of felspars by the Rb–Sr method is essentially confined to the K-felspars by the very unfavourable Rb/Sr ratios usually found in this group. Even K-felspars are frequently unsuitable because of this. Their radiogenic strontium retentivity is, however, apparently very good (see Table 4.7) and the factors responsible for the typical argon loss from felspars do not seem to disturb the Rb–Sr system.

AMPHIBOLES and PYROXENES

Extremely unfavourable Rb/Sr ratios occur in these minerals and scarcely a handful of data has been reported for them. A reasonable age of 950 m.y. was found for a hornblende by Pinson *et al.* (1958), while McDougall *et al.* (1963) incorporated one pyroxene point on an isochron drawn for British Guiana dolerites. In general, however, pyroxenes serve to indicate $(Sr^{87}/Sr^{86})_i$ values.

GLAUCONITES

Favourable Rb/Sr ratios in this mineral enable accurate "analytical ages" to be obtained (Cormier, 1956; Herzog *et al.*, 1958). Unfortunately, however, Rb–Sr ages on glauconites are often 10–20% below what should be comparable igneous mica ages. Having found an apparent correlation between the percentage of expandable layers, the potassium concentration and the age, Hurley *et al.* (1960) suggested that glauconites might take up "K and Rb in newly formed well-ordered additions, losing Ar^{40} and Sr^{87} from the expandable layer at the same time". In this way both K–Ar and Rb–Sr ages would be discrepantly low. Rb–Sr ages on glauconite, however, are apparently reliable indicators of the *minimum* age of a sediment. The correlation between potassium content and age for glauconites observed by Hurley *et al.* (1960) is shown in Table 4.8.

TABLE 4.8. VARIATION OF K-CONTENT WITH AGE OF GLAUCONITE
(After Hurley *et al.*)

Geologic age	%K Average	%K Range	No. of analyses
Lower Palaeozoic	6·1	4·9–6·9	11
Upper Palaeozoic and Mesozoic	5·1	3·3–6·2	9
Tertiary	4·1	2·3–5·8	6
Quaternary	3·0	—	1

4.7. Suitability of Whole Rocks for Rb–Sr Analysis

As we have already emphasized, the use of whole-rock samples dominates the scene in Rb–Sr geochronology. In an ideally undisturbed setting, separated minerals and whole-rock samples from igneous rocks will all define the same straight line on an isochron plot whose slope determines the age. Since minerals such as biotite will have more favourable Rb/Sr ratios it might be argued that undisturbed rocks might as well be dated by the use of minerals alone. In practice, however, this is not so, since minerals from rocks showing little or no visible effect of metamorphism or alteration may yet have been open systems on an atomic scale. The

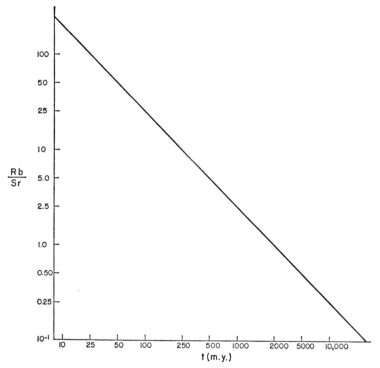

FIG. 4.4. Required Rb/Sr ratio to yield $Sr^{87}/Sr^{86} = 0 \cdot 8$, as a function of the age of the sample.

difficulty in finding whole-rock specimens with sufficiently high Rb/Sr values remains, however. In Fig. 4.4 we give a graphical version of equation (4.6) which shows, as a function of age, the Rb/Sr values required to make $(Sr^{87}/Sr^{86})_p = 0 \cdot 8$. In practice it is surprisingly difficult to find samples meeting this requirement, by far the most suitable rocks being granite or acid volcanic types. Basalts have such poor Rb/Sr values, generally, that for the most part they serve as indicators of common strontium composition.

4.8. Construction of the Rb–Sr Isochron

In drawing the Rb–Sr isochron through a set of experimental points one is faced with the task of finding the best straight line on a plot of y versus x when both y and x are subject to experimental errors. As this is a topic not well treated in most texts on statistics for experimentalists, a variety of least squares fittings have been used in isochron construction. It has been shown, however, by York (1966) and McIntyre (McIntyre *et al.*, 1966) that the slope of the best straight line is given by a root of the "Least Squares Cubic" equation in b

$$b^3 \sum_i \frac{W_i^2 U_i^2}{w(X_i)} - 2b^2 \sum_i \frac{W_i^2 U_i V_i}{w(X_i)} - b \left\{ \sum_i W_i U_i^2 - \sum_i \frac{W_i^2 V_i^2}{w(X_i)} \right\}$$

$$+ \sum_i W_i U_i V_i = 0, \tag{4.7}$$

where $U_i = X_i - \bar{X}$, $V_i = Y_i - \bar{Y}$, X_i and Y_i are the observed values,

$$\bar{X} = \frac{\Sigma_i W_i X_i}{\Sigma_i W_i}, \qquad \bar{Y} = \frac{\Sigma_i W_i Y_i}{\Sigma_i W_i},$$

$w(X_i)$ = the weight of X_i, $w(Y_i)$ = the weight of Y_i and

$$W_i = \frac{w(X_i)w(Y_i)}{b^2 w(Y_i) + w(X_i)}.$$

The weights $w(X_i)$ and $w(Y_i)$ are taken as the reciprocals of the variances of the X_i and Y_i respectively. The best intercept is given by

$$a = \bar{Y} - b\bar{X}. \tag{4.8}$$

The least squares cubic is not strictly a cubic equation in b since the terms in W_i involve b. However, if an approximate value for b is first substituted in the W_i, one obtains an exact cubic which may then be solved exactly in the usual fashion. The value for b thus obtained is now inserted in the W_i in place of the first approximate b and the cubic is re-solved. In this fashion the least squares cubic may be solved to any desired degree of accuracy. The properties of this equation are discussed in detail in York

(1966) where it is shown that the variances of the slope and intercept are given to a sufficient degree of accuracy by the expressions

$$\sigma_b^2 = \frac{1}{(n-2)} \frac{\Sigma_i W_i (bU_i - V_i)^2}{\Sigma_i W_i U_i^2} \tag{4.9}$$

and

$$\sigma_a^2 = \sigma_b^2 \frac{\Sigma_i W_i X_i^2}{\Sigma_i W_i}. \tag{4.10}$$

The best slope b minimizes the quantity S where

$$S = \Sigma_i \{w(X_i)(x_i - X_i)^2 + w(Y_i)(y_i - Y_i)^2\}.$$

The X_i and Y_i are the observables and x_i and y_i are the values that X_i and Y_i should have according to the least squares principle—they are the adjusted values of X_i and Y_i. The minimized value of S has a χ^2 distribution. On average, therefore its value should be about $n - 2$ (where n is the number of points plotted on the isochron) if the scatter of points about the line is consistent with the weights assigned on the basis of expected experimental errors in the X_i and Y_i. χ^2 tables will indicate whether or not a given value for S is highly improbable on the basis of the assigned weights. If the χ^2 tables indicate that the value found for S is only to be expected with a probability of less than 5%, then the implication is that (a) either the measurements are less accurate than supposed when the weights were calculated, or (b) the various assumptions underlying the expectation of a linear isochron plot have been violated to some extent. That is to say (a) some of the samples may have been open systems for Rb or Sr, (b) the samples may not have commenced with a common initial Sr^{87}/Sr^{86} ratio, (c) the samples may not, in fact, be of a common age. It is always important to examine the quantity S when an isochron is plotted. If S is satisfactory, then the slope calculated from the least squares cubic is immediately acceptable as indicating the common age of the samples. If S differs significantly from $n - 2$ because the isochron assumptions have been violated to some degree, then, strictly speaking, the age is uncertain. How to interpret the data, then, is largely a matter of individual judgement. It may be that one or two points cause the high value of S, in which case it might be reasonable to reject them and recalculate the slope. S can always be reduced by reducing the weights and new values for the slope, intercept and errors may be calculated. Various approaches of this nature are described in McIntyre et al. (1966).

The least squares cubic method is too long for reasonable hand calculation and standard programmes exist for its use with a computer. A graphical description of the way this method works has been given by York (1967) and this is summarized in Fig. 4.5.

Depending on the mode of experimentation adopted, there may be some correlation between the errors in Rb^{87}/Sr^{86} and Sr^{87}/Sr^{86}. When this is so a generalized version of the least squares cubic method of fitting should be used (York, 1969; Cumming, 1969).

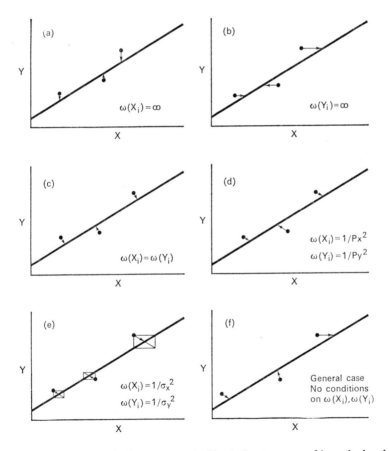

FIG. 4.5. Various types of adjustment required by the least squares cubic method under conditions of weighting stated on each graph. (After York, 1967.)

4.9. U–Pb and Th–Pb Methods

Natural uranium contains two radioactive isotopes, U^{238} and U^{235}, whose half-lives are of the same order of magnitude as the extent of geological time. An even longer lived isotope, Th^{232}, forms the bulk of natural thorium. The decays of these three parent isotopes to their respective stable end products, Pb^{206}, Pb^{207} and Pb^{208}, form the basis of three independent methods of age determination.

The decay processes proceed through three separate series of radioactive intermediate products, which, as shown in Tables 4.9, 4.10 and 4.11, have different half-lives as well as different chemical properties. The initial concentrations of these intermediates in a uranium–thorium mineral will depend to some degree on the geochemical processes which acted before and during the deposition of the mineral. If, because of these processes, a given intermediate is initially absent, its concentration will build up as its precursor decays. The product will itself decay at a rate which is always proportional to its atomic concentration. Its concentration will increase until eventually its rate of decay equals its rate of formation. After this condition of "secular equilibrium" is first reached, the concentration of the intermediate will be determined by its decay constant and the decay constants and concentrations of its longer lived precursors in the decay chain.

Eventually the most slowly decaying member of the chain determines the decay rate of all of the intermediate members. Under these circumstances the decay rate $N_i \lambda_i$ (equation (2.1)) of any intermediate member equals the decay rate $N_p \lambda_p$ of the parent isotope of the series, and the relative concentration of the intermediate N_i/N_p is therefore given by the ratio λ_p/λ_i. Because of the very long half-lives of the three parent isotopes, even the longest lived of the intermediate products have relatively low concentrations.

The intermediate decay products represent atoms which are held up in the process of transformation from uranium or thorium to lead. It might seem that for very young minerals, in which the amount of lead accumulated is small, failure to account for these intermediates might lead to serious errors in ages computed from the lead/uranium or lead/thorium ratios. In fact, if uranium minerals were formed initially free of U^{234} (the longest lived and most abundant intermediate in the U^{238} series), it would take several hundred million years before the total intermediate concentration decreased below $0 \cdot 1\%$ of the accumulated Pb^{206}. In *real* minerals, this effect is compensated by the presence of initial concentrations of the intermediates. Since U^{234} is an isotope of its parent, it is initially present in a uranium mineral in amounts close to its equilibrium abundance, and under these circumstances compensation is effected at a much earlier stage in the mineral's history.

For minerals older than a few million years, the numbers of atoms of Pb^{206}, Pb^{207} and Pb^{208} which accumulate in a time t are given with sufficient accuracy by the following equations in terms of the present amounts of U^{238}, U^{235} and Th^{232}:

$$Pb^{206} = U^{238} (e^{\lambda_{238} t} - 1), \qquad (4.11)$$

$$Pb^{207} = U^{235} (e^{\lambda_{235} t} - 1), \qquad (4.12)$$

and
$$Pb^{208} = Th^{232} (e^{\lambda_{232} t} - 1), \qquad (4.13)$$

and from these

$$t_{206} = \frac{1}{\lambda_{238}} \ln(1 + Pb^{206}/U^{238}), \qquad (4.14)$$

$$t_{207} = \frac{1}{\lambda_{235}} \ln(1 + Pb^{207}/U^{235}), \qquad (4.15)$$

and
$$t_{208} = \frac{1}{\lambda_{232}} \ln(1 + Pb^{208}/Th^{232}). \qquad (4.16)$$

An alternative (but not independent) method of computing an age arises as a result of the essentially identical properties of the two parent uranium isotopes. The equations for the production of radiogenic Pb^{206} and Pb^{207} may be combined to give

$$\frac{Pb^{207}}{Pb^{206}} = \frac{U^{235}}{U^{238}} \frac{(e^{\lambda_{235} t} - 1)}{(e^{\lambda_{238} t} - 1)}. \qquad (4.17)$$

The ratio U^{235}/U^{238} has a present value of $1/137 \cdot 8$, and has been shown experimentally to be independent of the age and history of the minerals from which the uranium was obtained. Using relation (4.17) it is possible to determine an age value for a mineral from a measurement of the ratio of radiogenic Pb^{207}/Pb^{206} alone. In early studies this method was attractive for two reasons. It avoided the necessity of quantitative uranium and lead determinations, and because it is based on a measurement of the ratio of two lead isotopes, the age computed is much less sensitive to losses of lead and uranium than ages calculated from isotopic Pb/U or Pb/Th ratios. Recent work has made it clear that the importance of the internal check which the Pb^{206}/U^{238} and Pb^{207}/U^{235} ratios provide far outweighs any advantage that might be derived from the use of the lead ratio method alone.

The problem of accounting for original concentrations of lead isotopes in uranium- and thorium-bearing minerals is similar to that encountered with the Rb–Sr method. Geochemical processes which segregate the parent isotopes are imperfect. Even though the chemical properties of uranium, thorium and lead are quite different in the earth's crust, small but often significant amounts of the daughter isotopes are captured with the parents at the time of formation of a rock or mineral. These initial concentrations must be subtracted from the total amount of daughter present before the ratio D/P is computed. The assumption is usually made that the original lead which contaminates minerals at the time of their formation has an isotopic composition similar to lead found in lead minerals of a corresponding age. In addition to the isotopes Pb^{206}, Pb^{207} and Pb^{208}, lead minerals contain small quantities (about $1-1 \cdot 5 \%$) of Pb^{204}, an isotope which is not produced by radioactive decay. This isotope provides a convenient index of the amount of lead incorporated initially in a uranium-bearing mineral, once the isotopic composition of the original lead is known.

The isotopic ratios Pb^{206}/Pb^{204}, Pb^{207}/Pb^{204} and Pb^{208}/Pb^{204} are rather strongly time dependent. These time dependences, shown in Fig. 6.1, are sufficiently well known that they contribute very little error to figures for original lead corrections. It is the low abundance of Pb^{204} and the difficulty in measuring its concentration accurately in the uranium or thorium mineral which determines the magnitude of the uncertainty.

The degree of concordancy among the U–Pb methods which may be achieved is shown in Table 5.1. The mineral most commonly used in U,Th–Pb dating is zircon. Also studied in some detail are uraninite, monazite and thorite. Analyses of such minerals show that U,Th–Pb ages are frequently discordant and it is usually necessary to interpret the data using the theories presented in Section 5.5.

4.10. Dating Minerals with Fossil Fission Tracks

Heavy charged particles moving through crystals leave trails of radiation-damaged material. Such tracks were originally examined by Silk and Barnes (1959). The width of the damaged region is about 100 Å and direct observation may only be made by electron microscopy. Price and Walker (1962) showed that a simple etching process suffices to make the tracks visible with an optical microscope. Immersion of micas in HF for intervals varying from 20 seconds to half an hour, depending on the type of mica, allows the acid to dissolve away the damaged regions and produce fine tubular holes which are readily seen by transmitted light in an optical microscope.

Such background tracks have been found in a variety of minerals (e.g. quartz, orthoclase, mica and zircon) by Fleischer and Price (1964). The tracks must be produced by heavy particles (atomic masses greater than about 30) with energies greater than 100 MeV, and it seems that those produced by the spontaneous fission of U^{238} are the most likely to be responsible. Assuming this to be so then we may calculate the age of a mineral in the following manner. Suppose a mica is freshly cleaved and then leached in HF. Those fission tracks which reached the cleavage surface will be attacked by the acid and hollowed out. The density of tracks now seen on the cleavage surface may be readily shown to be given by

$$\rho_s = 2(\lambda_f/\lambda) \, NC \, (\exp(\lambda t) - 1) \int_0^{R_0} q(z)dz, \qquad (4.18)$$

where $\lambda_f =$ the decay constant for spontaneous fission of U^{238},
$\lambda =$ the total decay constant for U^{238},
$N =$ the number of atoms/cm^3,
$C =$ the concentration of U^{238} in atoms/atom,
$q(z) =$ the fraction of fission events occurring in the depth interval $z, z + dz$ which traverse the cleavage surface and hence result in visible tracks on etching,
$R_0 =$ the range of the fission fragments.

The factor 2 occurs if we assume a symmetrical distribution of uranium about the internal potential cleavage surface. Hence,

$$t = \frac{1}{\lambda} \ln \left\{ 1 + \frac{\rho_s}{2NC} \cdot \frac{\lambda}{\lambda_F} \cdot \frac{1}{\int_0^{R_0} q(z)dz} \right\}. \qquad (4.19)$$

It merely remains to determine the uranium concentration NC and the factor

$$\int_0^{R_0} q(z)dz.$$

In fact we need only know the product of these two terms and this may be found in a straightforward fashion by irradiation of the sample with thermal neutrons. These induce fission of the U^{235} and the tracks of these events which intersect the same cleavage surface are newly revealed by fresh etching. We then have for the density of newly induced tracks

$$\rho_i = NC\left(\frac{U^{235}}{U^{238}}\right) n_t\, \sigma_t \left[\int_0^{R_0} q(z)dz\right]_{U^{235}}$$

or

$$NC\left[\int_0^{R_0} q(z)dz\right]_{U^{235}} = \rho_i \left(\frac{U^{238}}{U^{235}}\right)\frac{1}{n_t\sigma_t}. \tag{4.20}$$

If we set

$$\left[\int_0^{R_0} q(z)dz\right]_{U^{235}} = \int_0^{R_0} q(z)dz,$$

then equation (4.20) combined with equation (4.19) gives

$$t = \frac{1}{\lambda}\ln\left\{1 + \frac{\rho_s}{2\rho_i}\frac{\lambda}{\lambda_F}\left(\frac{U^{235}}{U^{238}}\right) n_t\, \sigma_t\right\}, \tag{4.21}$$

where U^{235}/U^{238} = present ratio of the uranium isotopes, n_t = thermal neutron dose in neutrons/cm^2 and σ_t = cross-section for fission of U^{235} by thermal neutrons. If we substitute the values $\lambda = 1\cdot54 \times 10^{-10}\,\mathrm{y}^{-1}$, $\lambda_F = 6\cdot9 \times 10^{-17}\,\mathrm{y}^{-1}$, $(U^{235}/U^{238}) = 7\cdot26 \times 10^{-3}$ and $\sigma_t = 5\cdot83 \times 10^{-22}\,\mathrm{cm}^2$ we have for t in years

$$t = 6\cdot49 \times 10^9\ \ln\left(1 + 4\cdot72 \times 10^{-18}\, n_t\frac{\rho_s}{\rho_i}\right). \tag{4.22}$$

In Table 4.12 the data of Fleischer *et al.* (1964) are shown. Fission track counts on zircons were made by these authors, boiling phosphoric acid

TABLE 4.9. THE DECAY SERIES OF URANIUM-238

Element	Nuclide	Mode of decay	Half-life
Uranium I	U^{238}	α	$4 \cdot 51 \times 10^9$ y
↓			
Uranium X 1	Th^{234}	β^-	$24 \cdot 10$ d
↓			
Uranium X 2	Pa^{234m}	β^-	$1 \cdot 175$ m
Uranium Z	Pa^{234}	β^-	$6 \cdot 66$ h
Uranium II	U^{234}	α	$2 \cdot 48 \times 10^5$ y
↓			
Ionium	Th^{230}	α	$8 \cdot 0 \times 10^4$ y
↓			
Radium	Ra^{236}	α	1622 y
↓			
Radium emanation, radon, niton	Rn^{222}	α	$3 \cdot 82$ d
↓			
Radium A	Po^{218}	α, β^-	$3 \cdot 05$ m
99.98% | 0.02%			
α ↓ ↓ β^-			
Radium B	Pb^{214}	β^-	$26 \cdot 8$ m
Astatine	At^{218}	α, β^-	$1 \cdot 5$–2 s
99·9% | 0·1%			
α ↓			
Radium C	Bi^{214}	β^-, α	$19 \cdot 7$ m
β^-			
0·04% | 99·96%			
α | β^- Radon	Rn^{218}	α	$0 \cdot 019$ s
Radium C′	Po^{214}	α	$1 \cdot 64 \times 10^{-4}$ s
Radium C″	Tl^{210}	β^-	$1 \cdot 32$ m
Radium D	Pb^{210}	β^-	$19 \cdot 4$ y
Radium E	Bi^{210}	β^-, α	$5 \cdot 01$ d
5×10^{-5}% | 99 + %			
α ↓			
Thallium β^-	Tl^{206}	β^-	$4 \cdot 19$ m
Radium F	Po^{210}	α	$138 \cdot 4$ d
Radium G	Pb^{206}	stable	—

TABLE 4.10. THE DECAY SERIES OF URANIUM-235

Element	Nuclide	Mode of decay	Half-life
Actinouranium ↓	U^{235}	α	$7 \cdot 1 \times 10^8$ y
Uranium Y ↓	Th^{231}	β^-	$26 \cdot 64$ h
Protactinium ↓	Pa^{231}	α	$3 \cdot 43 \times 10^4$ y
Actinium	Ac^{227}	β^-, α	$21 \cdot 6$ y
$1 \cdot 2\%$ | $98 \cdot 8\%$ $\alpha\downarrow$			
Actinium K | β^-	Fr^{223}	β^-, α	22 m
$\sim 6 \times 10^{-3}\%$ | $99 + \%$ Radioactinium $\alpha\downarrow$	Th^{227}	α	$18 \cdot 17$ d
Astatine | β^-	At^{219}	α, β^-	$0 \cdot 9$ m
$\sim 97\%$ | $\sim 3\%$ | Actinium X	Ra^{223}	α	$11 \cdot 68$ d
$\alpha\downarrow$ | β^- Bismuth	Bi^{215}	β^-	8 m
Actinon	Rn^{219}	α	$3 \cdot 92$ s
Actinium A	Po^{215}	α, β^-	$1 \cdot 83 \times 10^{-3}$ s
$99 + \%$ | $5 \times 10^{-4}\%$ $\alpha\downarrow$ Actinium B | β^-	Pb^{211}	β^-	$36 \cdot 1$ m
Astatine	At^{215}	α	$\sim 10^{-4}$ s
Actinium C	Bi^{211}	α, β^-	$2 \cdot 16$ m
$99 \cdot 7\%$ | $0 \cdot 3\%$ $\alpha\downarrow$ Actinium C″	Tl^{207}	β^-	$4 \cdot 79$ m
$\downarrow\beta^-$ Actinium C′	Po^{211}	α	$0 \cdot 52$ s
Actinium D	Pb^{207}	stable	—

being used as an etchant. The agreement found between fission-track ages and independent figures is extremely encouraging and a considerable upsurge of interest in this age method is anticipated. Much further testing of the various assumptions involved is needed. The time constant for natural track fading has still to be evaluated but it would seem to be greater than 10^{+9} y in some minerals. Track densities range from essentially zero to more than $5 \times 10^4/\text{cm}^2$ in micas and may be considerably higher in zircons. Fleischer et al. (1964) pointed out the interest in comparing fission-track ages with discordant U–Pb ages of zircons.

The stability of fossil fission tracks has been studied by Fleischer et al. (1965) and Storzer and Wagner (1969). The latter suggested that annealing experiments could be used to remove the effect of thermal events on fission track ages.

TABLE 4.11. THE DECAY SERIES OF THORIUM-232

Element	Nuclide	Mode of decay	Half-life
Thorium $\downarrow$	Th^{232}	α	$1 \cdot 39 \times 10^{10}$ y
Mesothorium I $\downarrow$	Ra^{228}	β^-	$6 \cdot 7$ y
Mesothorium II $\downarrow$	Ac^{228}	β^-	$6 \cdot 13$ h
Radiothorium $\downarrow$	Th^{228}	α	$1 \cdot 910$ y
Thorium X $\downarrow$	Ra^{224}	α	$3 \cdot 64$ d
Thorium emanation, thoron $\downarrow$	Rn^{220}	α	$51 \cdot 5$ s
Thorium A $\downarrow$	Po^{216}	α	$0 \cdot 158$ s
Thorium B $\downarrow$	Pb^{212}	β^-	$10 \cdot 64$ h
Thorium C	Bi^{212}	β^-, α	$60 \cdot 5$ m
Thorium C″	Tl^{208}	β^-	$3 \cdot 10$ m
Thorium C′	Po^{212}	α	$3 \cdot 04 \times 10^{-7}$s
Thorium D	Pb^{208}	stable	—

$36 \cdot 2 \% \quad | \quad 63 \cdot 8 \%$

$\alpha \downarrow \qquad \beta^-$

TABLE 4.12. AGES AND URANIUM CONTENTS OF ZIRCONS

(After Fleischer *et al.*)

Location	Source	Crystal size	Fission-track age, 10^6 y	Radioactive decay age, 10^6 y	Uranium conc., ppm
Chaone Mt., S. Nyasaland	T. W. Stern	2 crystals, each several mm diam.	(a) 133 ± 15 (b) 120 ± 40	138 ± 14 138 ± 14	13 0·6
N. Elkhorn Mts., Montana	T. W. Stern	2 crystals, each 5 g	78 ± 10	100 ± 15 (Pb-α) 76 ± 4 (K–Ar on associated biotite)	120
Jemez Mts., N. Mexico	T. W. Stern	4 crystals, each 5 g	$0·9 \pm 0·1$ (outside) $0·9 \pm 0·2$ (inside)	In progress (stratigraphic age in Pleistocene)	730 (outside) 100 (inside)

CHAPTER 5

DATING DISTURBED MINERALS AND ROCKS

5.1. Introduction

In the previous chapter we were concerned with the dating of rocks and minerals which had remained geochemically undisturbed since their formation. Under these circumstances, the ages obtained by all of the methods for a given sample will agree reasonably well within the limitations imposed by the uncertainties in half-lives. This is illustrated in Table 5.1. While it is not generally feasible, or even possible, to date a rock or

TABLE 5.1. REASONABLY CONCORDANT U–Pb, K–Ar AND Rb–Sr AGES

(After Wetherill *et al.*)

Locality	Pb^{206}/U^{238}	Pb^{207}/U^{235}	K–Ar	Rb–Sr
Portland, Conn., U.S.A.	268	266	265	251
Glastonbury, Conn., U.S.A.	251	255	259	257
Spruce Pine, N.C., U.S.A.	370	375	349	352
Branchville, Conn., U.S.A.	367	365	382	
Parry Sound, Ont., Can.	994	993	970	
Cardiff Twp., Ont., Can.	1020	1020	1000	970
Keystone, S. Dakota, U.S.A.	1580	1600	1520	1570
Viking Lake, Sask., Can.	1850	1880	1850	1852
Bikita, S. Rhodesia	2640	2670	2550	2519

Ages are in millions of years. The decay constant of Rb^{87} was taken as
$$\lambda = 1 \cdot 39 \times 10^{-11} \, y^{-1}.$$

mineral specimen by all five of the major decay processes now in use, it is often possible to compare the results obtained by at least two methods, such as the K–Ar and Rb–Sr or U^{238}–Pb^{206}, U^{235}–Pb^{207} and Th^{232}–Pb^{208}. Where the results of comparisons of this sort disagree, it is clear that some sort of transfer of materials into or out of the rock or mineral has taken place.

75

It has also become apparent from the number of published discordant ages that disturbances of this nature are far more common than was formerly realized. In many cases the transfer can be described as a diffusion process, resulting basically from the unequal distribution of the parent isotope among the various minerals within a rock. Diffusion in solids is extremely slow, on a laboratory time scale, but the process is favoured by the comparatively long lifetimes of the geological systems with which we are dealing. Diffusion is often strongly temperature dependent, and is aided by the increased temperatures which accompany regional or local metamorphism. But perhaps the most important factors controlling diffusion are the internal structure of the minerals in which the process takes place, and the nature of the bonding between the diffusing isotopes and the major constituents. For these reasons it is simplest to discuss the disturbing processes and the dating of the times of disturbance (where this is possible) for the K–Ar and Rb–Sr processes separately from the U,Th–Pb systems, even though the problem of determining the complete chronology of a rock mass will often depend on age measurements made by all the methods.

5.2. K–Ar and Rb–Sr Methods

The effect of a thermal disturbance on K–Ar and Rb–Sr systems is illustrated by Fig. 5.1, taken from a study by Hart (1964). The Eldora quartz monzonite stock in Colorado, U.S.A. was intruded approximately 54 m.y. ago into Precambrian gneisses and schists which are characterized by mineral ages in the range 1000–1300 m.y. Hart examined the ways in which the various mineral clocks of this Precambrian terrain were reset by the tertiary stock. If we consider first the results for the K–Ar systems we see that the K–Ar ages of biotites were severely lowered up to distances of over 1000 feet from the contact. Hornblende, on the other hand, was much less affected, retaining much the greatest part of its original argon at distances from the contact as small as 10 feet. A somewhat irregular behaviour was noticed with K-felspar, which we have already seen is prone to giving anomalously low K–Ar ages. Hart suggested that the irregularities in the felspar behaviour in Fig. 5.1 and the curious "double plateau" in the hornblende results were due to the argon in these minerals being in different "phases". The reasoning was made by analogy with the results of Amirkhanoff et al. (1961) which are discussed in Chapter 7.

The biotite Rb–Sr ages were strongly reduced in much the same way

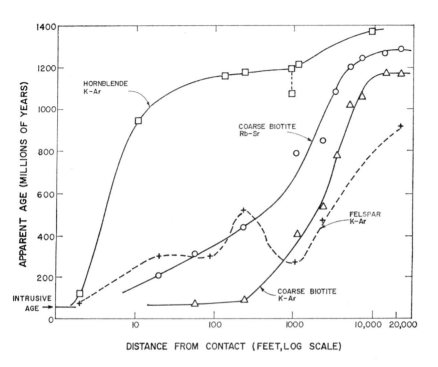

FIG. 5.1. Effect of a thermal disturbance on K–Ar and Rb–Sr systems. (After Hart, 1964.)

TABLE 5.2. ESTIMATES OF RESISTANCE OF AGE SYSTEMS TO THERMAL DISTURBANCE

Hart (1964)		Aldrich *et al.* (1965)
(a)	(b)	
		Zircon, 207/206
Felspar, Rb–Sr	Hornblende, K–Ar	Felspar, Muscovite, Rb–Sr
Hornblende, K–Ar	Felspar, Rb–Sr	Hornblende, K–Ar
	Felspar, K–Ar	Muscovite, K–Ar
Biotite, Rb–Sr	Biotite, Rb–Sr	Biotite, Rb–Sr
Biotite, K–Ar	Biotite, K–Ar	Biotite, K–Ar
Felspar, K–Ar		Felspar, K–Ar

as the K–Ar ages on the same mineral, except that the Rb–Sr values were systematically less affected. Rb–Sr systems for K-felspars, the data for which are not shown in Fig. 5.1, apparently were barely affected at all. From this study, Hart drew up the order of resistance to thermal disturbance shown in Table 5.2. In the same table we give the order of resistance estimated by Aldrich et al. (1965), who reported over one hundred independent age determinations on a variety of minerals from northern Michigan using all the major age methods where applicable. Their data indicated that, in a group of discordant results, the Pb^{207}/Pb^{206} ages for zircons usually yielded the greatest ages. After that their order is very similar to that of Hart. It would seem that (after the Pb^{207}/Pb^{206} ages) in many cases Rb–Sr ages on K-felspar and muscovite and the K–Ar ages of hornblendes are the most insensitive to change. The micas by both Rb–Sr and K–Ar seem to be one step further down the ladder. Potassium felspars are more difficult to position with regard to argon retentivity. In an undisturbed environment they regularly show spontaneous argon losses. On the other hand, they occasionally yield higher ages than micas if both have undergone a metamorphic episode. This effect is clearly seen in Fig. 5.1 and also in Fig. 4.2 as noted by Goldich et al. (1961).

There remain several interesting points to be made with regard to the micas. Firstly, the two sets of authors represented in Table 5.2 show the Rb–Sr ages of biotites being more resistant to thermal change than the K–Ar ages on the same minerals. Yet there are numerous cases in the literature of biotites, particularly metamorphic biotites, having higher ages by K–Ar than by Rb–Sr. To explain this phenomenon Kulp and Engels (1963) proposed that heating effects would always reset the K–Ar clock more effectively than the Rb–Sr clock, the degree of difference depending on the severity of the heating. Any instance of a biotite Rb–Sr age being less than the K–Ar age was then ascribed to a base–exchange effect due to circulating ground-waters. These authors suggested that such exchange would remove K^{40} and Ar^{40} equally effectively and thus cause no change in the K–Ar age. The Rb–Sr age, however, was considered to be lowered since rubidium, having a high chemisorption potential, would tend to stay in the mineral (it might even be gained), while the radiogenic Sr^{87} would leave the biotite in exchange for common strontium. Laboratory exchange experiments were found to support this hypothesis. In Table 5.3 from Kulp and Engels (1963) it may be seen how 50% of the potassium content of a biotite may be removed by exchange without

TABLE 5.3. EFFECT OF BASE EXCHANGE ON K–Ar
AGES

(After Kulp and Engels)

% Reduction in K conc.	K–Ar age (m.y.)
0	770 ± 20
7	740 ± 30
18	755 ± 25
19	830 ± 40
20	785 ± 25
53	805 ± 25
78	685 ± 20
87	695 ± 20
93	165 ± 5

any effect on the K–Ar age. Even the removal of 80 % of the original potassium caused a mere 10 % drop in age. These results are consistent with the fact that several investigators have found equal K–Ar ages on weathered and unaltered biotite samples (e.g. Baadsgaard *et al.*, 1961; Zartman, 1964). However, this mechanism for producing lower Rb–Sr biotite ages is not by any means finally accepted as being common, and the problem is exacerbated by the uncertainty of at least 6 % in the Rb^{87} half-life.

A second point of interest regarding the ordering of the micas in Table 5.2 is the position of muscovite K–Ar ages. Hart (1964) did not list muscovite and later has said (Hart, 1966) that there is little or no difference in argon retentivity among the various micas. In contrast with this, Aldrich *et al.* (1965) considered the muscovite K–Ar ages to be more resistant to change than both K–Ar and Rb–Sr ages on biotites, and this was a reasonable conclusion from their data. Further support for this viewpoint may be drawn from the results of Harper (1964) on coexisting biotites and muscovites. Harper noticed that the muscovite K–Ar age was *always* greater than that of the associated biotite by about 12 m.y. It might also be noticed that in Table 7.3 muscovite has clearly greater measured activation energies for argon diffusion than has biotite, which again would favour greater argon retentivity by muscovite. In several other studies, however, little or no difference has been observed between the argon retentivities of biotite and muscovite. In particular, Wanless and Lowden (1963) examined twenty-three biotite–muscovite pairs and found fourteen were concordant within the K–Ar experimental errors, five showed biotite below muscovite and four had muscovite below biotite. Ages ranged from 75 m.y. to 2·5 b.y. It is obvious that the situation is not simple. No doubt there are various processes in nature which conspire to produce reduced measured ages. One order of resistance may be consistently found in response to one process of disturbance while another order may be appropriate in a second characteristic type of perturbation. However, we are still some distance from resolving the problem into its various components, let alone assembling a final answer. A major barrier to be overcome now is erected by the uncertainties remaining in the half-lives of the various activities, with Rb^{87} being the chief offender and U^{235} running second.

5.3. Rb–Sr Whole-rock Method

The data $(Sr^{87}/Sr^{86})_p$, Rb^{87}/Sr^{86} obtained from biotites, muscovites and felspars extracted from the same *undisturbed* rock define a straight line on an isochron plot, as explained in Chapter 4. Furthermore, samples of the whole rock will supply points to fit the same line. However, an interesting experimental observation is that, following a metamorphism, the mineral points will have been disturbed from the straight line while in many cases the whole-rock points will not have been affected and will give the correct straight line (Compston and Jeffery, 1959; Nicolaysen, 1961). By taking the slope of the whole-rock line we get the age of original crystallization of the rock. Evidently, while movement of isotopes into and out of the various crystal phases can occur during metamorphism with the consequent production of mineral discordancies, the range of movement is small enough in a surprisingly large number of situations so that one can easily choose a sufficiently large volume of whole rock that has essentially remained a closed system. The concept is easily grasped when one considers the following argument. Suppose that during a metamorphic episode the average displacement of a radiogenic Sr^{87} atom is 1 cm. Then if we consider a whole-rock volume comprising a cube of sides 100 cm we see that the cube acts to an excellent approximation as a closed system. Almost all the movement associated with that volume takes place within its boundaries. Only a negligible transfer of material will occur into or out of the cube. As in nature the typical displacements are often much smaller than this, then the minimum size of the whole-rock specimen which may be regarded as a closed system during metamorphism is commensurately reduced.

It is not possible to over-emphasize the importance of the "whole-rock" concept of Rb–Sr dating. Where previously only inspired guesswork could be brought to play in the interpretation of discordant Rb–Sr mineral results, we can now in very many cases obtain experimentally the correct, unambiguous age of original formation of an igneous rock which has undergone at least one metamorphism before analysis. A good illustration of this is provided in a paper by Allsopp (1961). Table 5.4 shows the data

TABLE 5.4. DISCORDANT Rb–Sr MINERAL AGES FROM TRANSVAAL
(After Allsopp)

Source	Mineral	Apparent age (m.y.)
Half-way House granite	Felspar	3060 ± 100
	Biotite	2310 ± 40
	Chlorite	2890 ± 280
	Muscovite A	4540 ± 230
	Muscovite B	3820 ± 160
	Apatite	
	Epidote	
Witkoppen granite	Felspar	2580 ± 290
	Biotite	2120 ± 10
Half-way House pegmatite	Biotite	3010 ± 30
	Felspar	3000 ± 40
Witkoppen pegmatite	Felspar	2620 ± 60
Corlett Drive pegmatite	Felspar	2800 ± 30

for minerals extracted from samples of rock representative of the ancient Precambrian basement in the Transvaal area of South Africa. The ages of the minerals, calculated on the assumption that they trapped strontium at formation of composition $(Sr^{87}/Sr^{86})_i = 0 \cdot 71$, are grossly discordant. Similarly, as can be seen in Fig. 5.2, the mineral data on an isochron plot do not lie on any simple straight line. The whole-rock samples, however, accurately define a straight line, indicating that these whole rocks started from a common $(Sr^{87}/Sr^{86})_i$ value and that they have remained closed systems until they were examined in the laboratory. The slope of the line shows that the rocks crystallized originally 3200 ± 60 m.y. ago and ~~the~~ intercept on the ordinate gives $(Sr^{87}/Sr^{86})_i = 0.70$. Unfortunately we can say nothing very useful about just how and when the minerals became open systems because their distribution on the isochron diagram seems irregular.

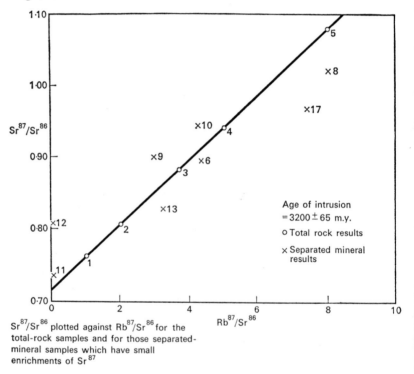

Sr^{87}/Sr^{86} plotted against Rb^{87}/Sr^{86} for the total-rock samples and for those separated-mineral samples which have small enrichments of Sr^{87}

FIG. 5.2. Rb–Sr isochron plot. (After Allsopp, 1961.)

Since the whole-rock samples enable us to see through the discordancy inflicted on the mineral data by a metamorphic event, so that we can easily calculate the time of original formation of a rock, it is natural to wonder whether we can sometimes in some way estimate the time at which the later disturbing event took place. Often this is not possible, but there is one particular case in which interpretation is a simple matter. Let us imagine the metamorphic episode is such that it causes the homogenization of the Sr^{87}/Sr^{86} ratio among *all* the various mineral phases *within a given whole-rock volume*. Then *at this time* all the minerals *in this volume* will have the same Sr^{87}/Sr^{86} value as the whole rock within which they are contained. The minerals in a different volume of the granite are supposed to be homogenized to *their* local whole-rock Sr^{87}/Sr^{86} value. Thus one envisages the rock mass to be composed of a multitude of discrete "whole-rock volumes" within each one of which the Sr^{87}/Sr^{86} isotope ratios are made homogeneous by the metamorphism. Clearly a present-day analysis of a rock which has undergone just such a history will yield Fig. 5.3. The whole-rock points will define a straight line whose slope gives the age of intrusion of the granite. They will not have "seen" the metamorphism. But the interesting thing now is that the various mineral points define a single set of parallel straight lines. The average slope of these lines gives the age of metamorphism. The other interesting observation is that the points of minerals from one particular whole-rock volume together with that whole-rock point fall on one line, other minerals and their whole-rock point fall on another, and so on. The intercept on the Sr^{87}/Sr^{86} axis of any mineral line gives the value of the Sr^{87}/Sr^{86} ratio to which those minerals were homogenized at metamorphism. Compston and Jeffery (1959) gave evidence of one occurrence of such a type of metamorphism. Several other examples also exist in the literature (see Fig. 5.4). However, just how many metamorphic episodes will be adequately described by such a model remains to be seen. If the local homogenization process described above did not occur in a metamorphism then the mineral points will not lie on sets of parallel lines but will be scattered on the diagram as in Fig. 5.2.

Lanphere, Wasserburg, Albee and Tilton (1964) gave an example where the metamorphism was such that even the whole-rock approach broke down.

5.4. $(Sr^{87}/Sr^{86})_i$

An extremely interesting fact, emphasized by Gast (1960), which has become more apparent with the increase in the number of whole-rock isochrons, is that the parameter $(Sr^{87}/Sr^{86})_i$ is remarkably constant for rocks of a wide variety of type and age. The obvious meaning of this observation is that the vast majority of igneous rocks have been formed from sources in which the Rb/Sr ratio is very low. It seems (Hurley *et al.*, 1962) that in a very large number of cases $(Sr^{87}/Sr^{86})_i \cong 0 \cdot 71$. For a mineral like biotite the assumption of such a figure is probably valid,

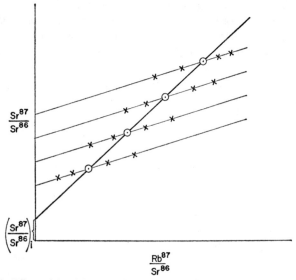

FIG. 5.3. Effect of local homogenization on Rb–Sr isochron plot. ⊙ = whole rocks; × = minerals.

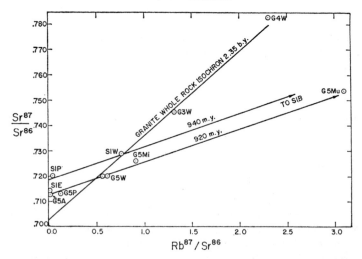

FIG. 5.4. Natural example of local homogenization. Whole rocks give an isochron age of 2·35 b.y., while separated minerals indicate local homogenization about 930 m.y. ago. (After Grant, 1965.)

provided the rock containing the mineral has been undisturbed. However, from what has been said above it is clear that if the rock has been disturbed, then the biotite age cannot be found by adopting such a value for $(Sr^{87}/Sr^{86})_i$ unless the Rb/Sr ratio is very high. Since there is no infallible criterion for deciding, before an isotopic analysis, whether or not a mineral has gained or lost Sr^{87} by diffusion, it is evidently extremely ill-advised to attempt to date a rock by simply carrying out analyses on a separated mineral fraction and assuming a value of $(Sr^{87}/Sr^{86})_i$ equal to $0·71$. The dating of old metamorphic shield areas by the Rb–Sr method must be carried out by whole-rock dating in conjunction with analyses of the various separated minerals. Only in this way will such of the ambiguity in the interpretation of presently existing mineral data be removed.

5.5. U–Pb and Th–Pb Methods

The majority of the many published uranium–lead and thorium–lead age determinations are discordant. As a result a number of studies have been made in attempts to determine how these discordancies are produced and how the discordant ages are related to the true age of crystallization.

A number of typical discordant ages are given in Table 5.5. It is evident

TABLE 5.5. DISCORDANT U–Pb AND Th–Pb AGES

Sample	206/238	207/235	207/206	208/232	Refs.
Swedish Kolm (S-9K)	365	435	920		1
Uraninite, Huron Claim, S.E. Manitoba	1564	1985	2475	1273	2
Zircon, Quartz Creek, Colorado Granite A.	925	1130	1540	530	3
Zircon, Johny Lyon Granodiorite, Arizona (R200 mesh)	1070	1270	1630	1660	4

1. Cobb and Kulp (1961). 3. Tilton *et al.* (1957).
2. Nier (1939). 4. Silver and Deutsch (1963).

All ages in m.y.

that two Th–Pb ages are lower than the U–Pb ages. This is a fairly general finding, even when the latter are nearly concordant, and it suggests that in minerals containing both uranium and thorium, the thorium is much more easily removed by chemical processes than the uranium. Laboratory tests made by Tilton (1956), in which specimens were leached with acid, have demonstrated this effect, and for this reason ages based solely on Pb/Th ratios are generally suspect.

The U–Pb ages in Table 5.5 exhibit a regular pattern which is typical of almost all discordant ages determined by this method. For each sample the Pb^{206}/U^{238} age is the lowest, the Pb^{207}/U^{235} age is somewhat higher, and the Pb^{207}/Pb^{206} age is the highest. This relationship was noted by several of the early workers in the field of age determinations, and its recognition led to two quite different interpretations. Wickman (1942) suggested that the discrepancies in the ages determined for the Swedish Kolm samples (see Table 5.5) might be due to the preferential loss of one of the intermediate products in the U^{238} decay series. Because of their low concentrations a *sudden* loss of any or all of the intermediates will not appreciably affect ages computed from the Pb/U ratio. On the other hand, a *continuous* loss over the lifetime of a mineral of a fraction of any intermediate will result in an equivalent percentage decrease in the concentration of the resultant stable end product.

Continued loss of this sort is not inconceivable for minerals containing uranium and thorium. During the decay of the radioactive isotopes in all of the decay series relatively large amounts of energy are released. This appears mainly as kinetic energy of the alpha and beta particles released in the decay processes, as recoil energy of the decay product nuclei, and as energetic gamma rays. Much of the energy in these forms is absorbed by the crystal lattice of the mineral, and gives rise to dislocations and imperfections which ultimately result in the formation of minute cracks, if the concentration of the radioactive elements is sufficiently high. This internal radiation damage tends to provide paths whereby material may be transferred into or out of the mineral with greater facility than in the undamaged state.

The intermediate which seemed to Wickman to be most likely to escape was the rare gas isotope Rn^{222}. The much longer half-life of $3 \cdot 8$ day Rn^{222}, in the U^{238} decay series, seemed to favour its escape over that of $3 \cdot 9$ second Rn^{219} in the U^{235} series. Continuous loss of Rn^{222} by diffusion would result in a reduced concentration of Pb^{206} in the mineral, and so lower the Pb^{206}/U^{238} ratio and the age computed from it. The Pb^{207}/U^{235} age would be unaffected by this loss mechanism, while the Pb^{207}/Pb^{206} age would be increased.

On the other hand, Nier (1939), who made the first accurate isotopic analyses of lead in uranium minerals, pointed out that for several of his samples loss of lead during alteration or leaching at some instant during the lifetime of the minerals could produce discordant ages. In this case the discordancies are not due to *geochemical* differences, but result from the non-linear character of expressions (4.14)–(4.16) for the age of the mineral, coupled with the fact that the decay constants of the two parent isotopes are different. Loss of lead will produce in this way the pattern of discordant ages observed in Table 5.5.

There were sound geological and physical reasons for accepting these two quite different mechanisms in the individual cases for which they were proposed. Since the two mechanisms led to very different estimates of the true age of formation, it was essential to determine whether one or the other described the real loss process acting in other discordant systems. Laboratory measurements using fresh uranium minerals have since suggested that radon loss cannot explain the magnitude of most of the observed discordancies (Giletti and Kulp, 1955).

It seemed that loss of lead was the most likely explanation, but no real progress toward a proper understanding of the mechanism was made until the work of Ahrens (1955) and in particular, Wetherill (1956). The latter showed that if one plotted Pb^{206}/U^{238} against Pb^{207}/U^{235} for concordant samples of various ages, the points should define a single curve, which Wetherill named "concordia". This concordia plot is shown in Fig. 5.5. The relation of discordant lead–uranium ratios to concordia can be seen by considering the effect of removing at the present time varying amounts of lead from a concordant mineral of some age t_1. If the process does not discriminate between Pb^{207} and Pb^{206}, these isotopes will always be removed in equal proportions. Consequently the resulting Pb^{206}/U^{238} and Pb^{207}/U^{235} ratios will lie along a straight line joining point t_1 on concordia to the origin. The position of any given point on the line will, of course, depend on the fraction of the lead extracted from the sample. The same reasoning applies to lead extracted from a mineral by some geochemical process at a time t_2 in the past. Clearly, if the concordia plot had been constructed t_2 years ago, its origin would have been at the point which is now t_2 on the present-day concordia. The Pb^{206}/U^{238} and Pb^{207}/U^{235}

in the discordant sample would lie on a line joining t_1 and t_2, the exact position depending still on the fraction of lead removed.

A suite of samples all of the same original age t_1, which lost various amounts of lead at a time t_2 years ago, would therefore be expected to have Pb^{206}/U^{238} and Pb^{207}/U^{235} ratios which defined a straight line intersecting concordia at times t_1 and t_2 as illustrated in Fig. 5.5. The slope and position of the line are in fact functions only of the times t_1 and t_2, and loss or

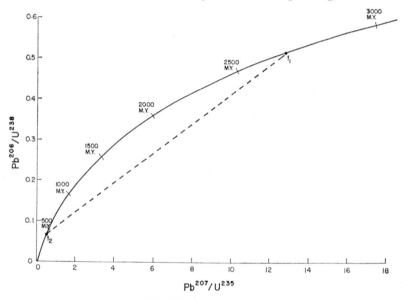

FIG. 5.5. Concordia plot.

gain of uranium (which might well be expected to occur during a disturbing event) simply changes the position of a point on the line. This interpretation provides an estimate for the time at which the loss of lead took place as well as a measure of the true age of crystallization.

Most of the suites of discordant Pb/U ratios which have been published to date do define straight lines on the concordia plot. A number of these are shown in Fig. 5.6. In general the true ages of the deposits obtained from the intersections of the lines with concordia seem reasonable when compared with ages determined by other methods. But one factor delayed the acceptance of the loss mechanism which the existence of these linear relationships inferred. This was the curious one-to-one relationship between

t_1 and t_2, which is quite apparent in Fig. 5.6. In addition discordant uranium minerals from the rocks of ancient continental nuclei in several continents lie on the same line. There is good evidence that these rocks shared a roughly common origin in time, but it is difficult to accept the apparent conclusion from the Pb/U data that they were all disturbed at the same time roughly 600 m.y. ago. The difficulty led several investigators to consider the possible consequences of other loss mechanisms.

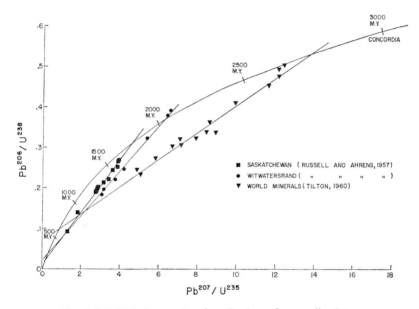

FIG. 5.6. Published examples of applications of concordia plots.

The apparent relation between times t_1 and t_2 could be produced by a process which favoured the removal of Pb^{207} relative to Pb^{206}. No mechanism for removing lead *at a single instant* seems capable of discriminating between lead isotopes in this way. It is necessary to consider continuous processes in which either intermediate decay products or radiogenic lead are lost over the entire lifetime of the mineral.

One such process involves the recoil energies of those intermediates which decay by alpha particle emission. More energy is released by recoil in the U^{235} decay series than in the U^{238} chain. Russell and Ahrens (1957) suggested in a qualitative manner that if this recoil process led in any way to the loss of intermediate products or lead, the net result would be an effectively larger loss of Pb^{207} than of Pb^{206}.

An alternative proposal which has been considered in some detail, is loss by volume diffusion, not of intermediate decay products, but of lead. The initial suggestion and computations were made by Nicolaysen (1957), but the application of the concept to the specific purpose of explaining the distribution of discordant Pb/U ratios on the concordia plot was made by Tilton (1960). Tilton demonstrated that the continuous diffusion of radiogenic lead from a zircon crystal of age t_1 would result in Pb/U ratios which lay on a curve passing through t_1. Over the greater part of its length the curve was nearly linear, and matched extremely well the experimentally defined line. The position of the point on the curve was defined by a parameter D/a^2, the diffusion coefficient divided by the square of the effective radius of the crystal.

It should be noted that the preferential removal by diffusion of Pb^{207} relative to Pb^{206} is not due to any fundamental difference in the property of the lead isotopes themselves. The effect is due solely to the different half-lives of the respective parent uranium isotopes. Because of the shorter half-life of U^{235}, Pb^{207} atoms in a given mineral are *on the average* somewhat older than Pb^{206} atoms. The Pb^{207} atoms will therefore tend to diffuse out of the mineral to a greater degree than the Pb^{206} atoms because they have *on the average* a greater length of time available for this process.

The fit of the theoretical diffusion loss curve to several groups of experimental data is given in Fig. 5.7. A mineral which had lost all of its lead by continuous diffusion would now have Pb/U ratios lying at the origin of concordia. The theoretical diffusion loss curves must therefore pass through the origin. It is apparent that if the diffusion mechanism correctly describes the manner in which discordant U–Pb ages are produced, the intercept t_2 is not real. As can be seen in Fig. 5.7 the plot is strongly curved only at high values of D/a^2, in a region remote from any of the experimental points yet available. Since it is in this region that the prediction of the diffusion loss and chemical loss mechanisms differ, it is here that critical experiments to decide between them would have to be made. This would necessitate finding and analysing crystals to which very large values of D/a^2 applied.

Although no laboratory studies of the diffusion of lead through zircon have been made, the D/a^2 values for real minerals as estimated from the concordia plot appear to be of the correct order of magnitude.

The data also suggest that the diffusion coefficient is relatively insensitive to temperature changes. To explain this, Wasserburg (1963) has suggested that the internal radiation damage mentioned earlier might be the factor which provides the diffusion paths for lead loss. Since the damage is cumulative one might expect the diffusion coefficient to be an increasing

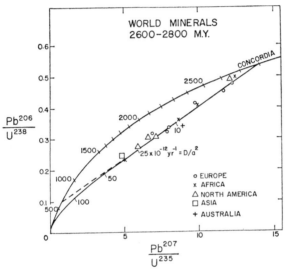

Fig. 5.7. Continuous lead-diffusion locus, effectively straight over much of its length, curving eventually to the origin. (After Tilton, 1960.)

function of time. A linearly time-dependent diffusion coefficient leads to theoretical diffusion loss curves which fit the existing data as well as, if not better than, the curves for constant coefficients. More accurate measurements of Pb/U isotopic ratios are necessary to decide this issue.

Regardless of which of the mechanisms applies in a given case, it appears that the true age of formation can be estimated for a suite of samples if their discordant Pb/U ratios produce a linear pattern when plotted on a concordia plot. This fact has already given new impetus to the determination of the ages of acid igneous rocks by the lead–uranium method. Thus Silver and Deutsch (1963) showed that zircons extracted from a 250 pound mass of granitic rock could be grouped into several families, on the basis of radioactivity and other properties, which yielded a linear array on a concordia plot. The indications were that this enabled very accurate estimates to be made both of the time of crystallization of the intrusive and of a subsequent metamorphism. These investigations also emphasized the importance of the uranothorite impurity.

5.6. Whole-rock U–Pb, Th–Pb Dating

With the introduction of whole-rock Rb–Sr dating it became apparent that the identical approach could in principle be adopted with the U–Pb and Th–Pb systems. By analogy with equation (4.3) we may write

$$\left(\frac{Pb^{206}}{Pb^{204}}\right)_p - \left(\frac{Pb^{206}}{Pb^{204}}\right)_i = \frac{U^{238}}{Pb^{204}}\,(e^{\lambda t} - 1) \tag{5.1}$$

where λ is the decay constant of U^{238}. Similar equations may be written for the U^{235}–Pb^{207} and Th^{232}–Pb^{208} systems. One is thus presented with the possibility of three whole-rock diagrams on which are plotted

$$\frac{Pb^{206}}{Pb^{204}} \text{ vs. } \frac{U^{238}}{Pb^{204}}, \quad \frac{Pb^{207}}{Pb^{204}} \text{ vs. } \frac{U^{235}}{Pb^{204}} \text{ and } \frac{Pb^{208}}{Pb^{204}} \text{ vs. } \frac{Th^{232}}{Pb^{204}},$$

respectively. By exact analogy with the reasoning for Rb–Sr whole-rock analysis, the ages would be calculated from the slopes of any resultant straight lines and the intercepts on the lead ratio axes would give the initial isotope ratios of the lead in the systems. This approach was discussed by Nicolaysen (1961) and applied to the data of Zartman (1965) by Ulrych and Reynolds (1966). It is to be expected that this method will receive increasing attention in coming years.

General reviews covering many of the topics in this chapter and Chapter 4 are given by Rankama (1954, 1963), Hamilton *et al* (1962), Hamilton (1965), Moorbath (1965) and Hart (1966).

CHAPTER 6

LEAD ISOTOPE METHODS

6.1. Model Ages

It was mentioned earlier, in connection with corrections for original lead in uranium minerals, that the ratios Pb^{206}/Pb^{204} and Pb^{207}/Pb^{204} in lead minerals are related in a general way to the age of formation. Isotopic equilibrium fractionation of lead isotopes is quite incapable of producing changes of the size and kind observed. The isotopic variations shown in Fig. 6.1 must therefore be related to additions of radiogenic lead from the decay of uranium and thorium.

The time dependence of the isotope ratios is inverse in the sense that older minerals generally contain *smaller* amounts of the radiogenic isotopes Pb^{206}, Pb^{207} (and Pb^{208}) than younger ones. Since lead minerals contain negligible amounts of uranium and thorium, changes in isotopic composition can only have been produced *prior* to the time of final deposition. In contrast to uranium minerals, then, the isotopic ratios in a lead mineral are affected by, and hence contain information about, events which occurred before the mineral's final emplacement.

Whatever geochemical processes individual leads may have undergone, it is clear from Fig. 6.1 that some common and world-wide process has controlled the main trend of the lead isotope abundance variations. Attempts to fit the observed pattern to a simple geological model were first made by Gerling (1942), Holmes (1947), and Houtermans (1947). In this model, it was assumed that at a time t_0 in the past, lead having some initial or primeval isotopic composition $a_0 = (Pb^{206}/Pb^{204})_0$, $b_0 = (Pb^{207}/Pb^{204})_0$ and $c_0 = (Pb^{208}/Pb^{204})_0$ was introduced into one or more closed rock systems with uranium and thorium. If the uranium and thorium concentrations were of the same order of magnitude as that of the lead, then the decay of these parent isotopes would gradually significantly alter the overall isotopic composition of the lead in the following ways:

93

$$\frac{Pb^{206}}{Pb^{204}} = a_0 + \frac{U^{238}}{Pb^{204}} \left(e^{\lambda_{238} t_0} - e^{\lambda_{238} t} \right), \tag{6.1}$$

$$\frac{Pb^{207}}{Pb^{204}} = b_0 + \frac{U^{235}}{Pb^{204}} \left(e^{\lambda_{235} t_0} - e^{\lambda_{235} t} \right), \tag{6.2}$$

$$\frac{Pb^{208}}{Pb^{204}} = c_0 + \frac{Th^{232}}{Pb^{204}} \left(e^{\lambda_{232} t_0} - e^{\lambda_{232} t} \right). \tag{6.3}$$

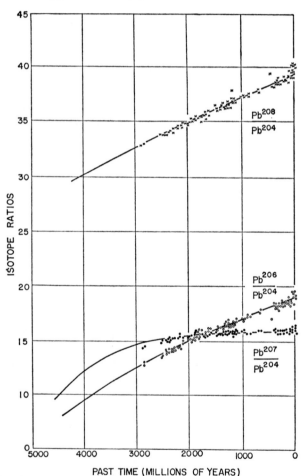

FIG. 6.1. Common lead isotope ratios as a function of the age of the mineral.
(After Russell and Farquhar, 1960.)

In these equations, Pb^{206}/Pb^{204}, Pb^{207}/Pb^{204} and Pb^{208}/Pb^{204} are the lead isotope ratios t years ago in a system in which the present-day ratios of uranium and thorium to Pb^{204} are U^{238}/Pb^{204}, U^{235}/Pb^{204} and Th^{232}/Pb^{204}. A convenient way of visualizing some of the important features of lead isotope variations is to plot Pb^{207}/Pb^{204} against Pb^{206}/Pb^{204}. As shown in Fig. 6.2, the time-dependent equations combine to give a family

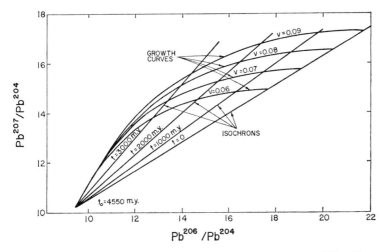

FIG. 6.2. Theoretical growth curves for lead isotope ratios. $V = U^{235}/Pb^{204}$.

of *growth curves*, each one of which is specified by a particular value of the parameter U^{238}/Pb^{204}. Each of these curves represents the course along which lead isotope ratios would evolve in a system presently containing a given U^{238}/Pb^{204} ratio. Equations (6.1) and (6.2) may also be combined to give

$$\frac{Pb^{207}/Pb^{204} - b_0}{Pb^{206}/Pb^{204} - a_0} = \frac{U^{235}}{U^{238}} \frac{(e^{\lambda_{235}t_0} - e^{\lambda_{235}t})}{(e^{\lambda_{238}t_0} - e^{\lambda_{238}t})}. \qquad (6.4)$$

Since the right-hand side of this equation is a function only of time t, systems of the same age but different U^{238}/Pb^{204} ratios will contain lead isotope ratios lying along straight lines, termed *isochrons*. These isochrons will diverge from the assumed common lead isotope ratios a_0, b_0 (Fig. 6.2) and will have slopes which are a function of t_0 and the age t.

Now any geochemical process which extracts and concentrates a representative sample of the lead in a given system into a lead mineral provides us with a record of the isotopic composition of the lead in that system at the time of extraction. If we are willing to accept the above model, therefore, and can evaluate the parameters in the equations, it should be possible to date this extraction event for a lead mineral whose Pb^{207}/Pb^{204} and Pb^{206}/Pb^{204} ratios are known, merely by substituting these ratios in equation (6.4) and solving for t. The values of the primeval lead isotope ratios were defined rather decisively by Patterson's analyses of lead in iron meteorites ($a_0 = 9 \cdot 56$, $b_0 = 10 \cdot 42$). The initial time t_0 is given the value of $4 \cdot 55$ b.y. which Patterson showed could reasonably be applied to the earth as well as to the time of uranium–lead fractionation in the stone meteorites (see Chapters 11 and 12).

6.2. Anomalous Leads

Acceptance of the model has, however, come rather more gradually, and only in a rather restricted sense. It was realized, even when a relatively small number of isotopic analyses were available, that there were some lead minerals having exceptional isotopic compositions which could not be described by the simple theory. These leads had isotopic ratios which lay to the right of the zero isochron in Fig. 6.2. Since a more complicated model was evidently required to explain them, they were termed *anomalous* or *J-type* leads (after Joplin, Missouri, where the first examples were found).

Further analyses of lead minerals from areas containing anomalous leads have revealed another characteristic of this class. In a given area the lead isotope ratios vary from sample to sample and the Pb^{206}/Pb^{204} ratios are linearly related to the Pb^{207}/Pb^{204} ratios. Three examples of this relationship are shown in Fig. 6.3. The simplest explanation of these linear trends is that they represent "secondary" isochrons. In each case some "parent" lead having an isotopic composition at or to the left of the left end of each sequence has had added to it varying amounts of radiogenic lead which have accumulated in surrounding rocks from the decay of uranium and thorium. In rocks of a given age, the radiogenic Pb^{207}/Pb^{206} ratio at any subsequent time will be everywhere the same. It follows that the isotopic ratios of the resulting mixtures of this radiogenic lead and the "parent" lead must lie on a straight line on a Pb^{207}/Pb^{204}–Pb^{206}/Pb^{204} graph (Russell and Farquhar, 1960).

Very little is known about the exact nature of the event in which addi-

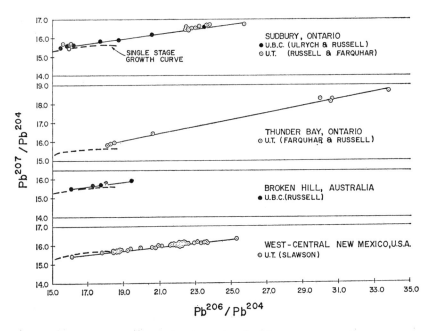

FIG. 6.3. Anomalous lead lines.

tions of this sort take place, but it is certainly the local scale of the process which preserves the heterogeneities in isotopic composition. From Fig. 6.3 it is evident that the fraction of radiogenic lead added to the "parent" lead varies over a wide range. If the fraction is large enough, the resulting lead minerals will be clearly anomalous. On the other hand, small additions of radiogenic lead could produce leads whose isotopic compositions might lie to the left of the zero isochron. One of these samples, if not accompanied by further analyses, could erroneously be "dated", on the assumption that the simple model applied to it.

Anomalous leads, once recognized, can be dealt with as a separate class of samples. The question arises as to whether there is any criterion whereby non-anomalous leads can be recognized. Stanton and Russell (1959) have suggested that sulphide deposits conformable with the rocks in which they lie might have had rather simple histories, and that any lead in them might therefore be expected to have had little chance of becoming anomalous. In fact there appear to be no detectable isotopic variations within those deposits of this type which have been sampled. As Fig. 6.4 indicates, the lead isotope ratios for all these deposits lie close to (but not exactly on) a single growth curve. This remarkable consistency suggests that the history of conformable leads may in fact be fairly well represented by the Holmes–Houtermans model, and that their ages may be estimated on this basis.

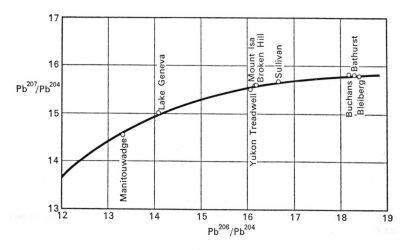

FIG. 6.4. Conformable lead growth curve. (After Russell and Farquhar, 1960.)

Attempts have also been made to utilize for dating purposes the linear isotopic relationships found for suites of anomalous lead minerals (Russell and Farquhar, 1960; Kanasewich, 1962). If the radiogenic lead which has been incorporated into the anomalous leads has developed in rocks whose age t_1 is known, then this age, together with the measured slope (the radiogenic Pb^{207}/Pb^{206} ratio $= R$) gives us sufficient information to calculate the time t_2 at which the anomalous leads were deposited. The equation to be solved is

$$ R = \frac{1}{137 \cdot 8} \left\{ \frac{e^{\lambda_{235} t_1} - e^{\lambda_{235} t_2}}{e^{\lambda_{238} t_1} - e^{\lambda_{238} t_2}} \right\}. \tag{6.5} $$

These lead isotope dating methods suffer from the difficulty of validating the calculated ages. No method of directly dating lead minerals is available.

It may be possible to check the accuracy of the Holmes–Houtermans model for other minerals such as pegmatitic potassium felspars, which contain lead as a minor constituent, but the relationship of these to lead minerals is not known. For reviews of the development of this topic reference should be made to Russell and Farquhar (1960) and Patterson (1964).

CHAPTER 7

THE GENERAL PROBLEM OF INTERPRETATION

7.1. Introduction

As data have steadily accrued, the problems involved in the interpretation of radiometric ages have become more clearly focused. A central problem which is rapidly emerging is that of the effect of long cooling histories on the ages recorded by the various systems. Two somewhat distinct schools of thought seem to exist in relation to this. On the one hand, it is argued that, in general, the radiogenic daughters are retained fully by the host crystals very shortly after the time of crystallization. That is, it is suggested that the age obtained indicates the time of crystallization, within the accuracy of the measurement. On the other hand, it is maintained that very many rocks, being formed during orogeny at great depth, take many millions of years to cool to temperatures at which daughter retention is essentially perfect.

Such a debate becomes very significant when the tectonic and metamorphic history of a complex shield area is being studied geochronometrically. Thus it is an experimental fact that the Grenville Province of the Canadian shield is characterized by minerals yielding ages by the U–Pb, Rb–Sr and K–Ar methods in the very wide range 800–1200 m.y. According to the first school of thought one could argue that the spread is due to the influence of several sharp metamorphic episodes and such fine structure as exists would be evaluated along those lines. Following the second hypothesis one could say that while several metamorphic episodes no doubt took place, the times of occurrence of these might be very difficult to define because of the long-continuing daughter diffusion in a deeply buried, active orogen.

The use of the slow-cooling hypothesis in geochronology is recent and it is interesting to consider the points favouring slow cooling of orogenic belts and the consequently prolonged daughter diffusion.

7.2. Slow-cooling Hypothesis

Hurley *et al.* (1962) reported K–Ar and Rb–Sr measurements on biotites from mica schists in south-west New Zealand. On geological grounds it was considered that these minerals probably formed at least 150 m.y. ago during metamorphism in the Jurassic. K–Ar ages for these minerals, however, fell between 4 and 8 m.y. A similar mica age value was obtained with the Rb–Sr method. The schists have been revealed by recent faulting and uplift. In explanation of the anomaly, Hurley *et al.* suggest that the schists were buried at a depth of approximately 3 km, for over 100 m.y. During this time the schists were supposed to be at about 100°C, which temperature is postulated to be high enough to cause complete degassing of the micas. Only as a result of the recent uplift were the schists able to cool to a temperature at which radiogenic daughter isotopes are retained. On this interpretation, then, the low ages actually date the time of uplift and erosion in this area. It is worth quoting the stimulating conclusions drawn by Hurley *et al.*

> Geological history has been recorded in terms of episodes of sedimentation because of palaeontological correlation. Owing to the difficulty of dating sediments the geochronologists have been making a plea for a swing toward orogenic and volcanic episodes as a frame of reference for the extension of earth history into the Pre-Cambrian. If the above conclusions are true, the ages commonly measured by K–Ar and Rb–Sr on biotite may actually reflect the time of major uplift and erosion, which coincides with that of sedimentation, and not the initial period of metamorphism and igneous intrusion in the orogenic belt. For example, in New Zealand the biotite age values post-dated the time of metamorphism by more than 100 m.y. Thus the geochronologists may be actually dating episodes that are more nearly comparable to times of sedimentation than to metamorphism, and the frame of reference provided by each method of dating may be more nearly matching than expected.

More recently Armstrong *et al.* (1966) have reported some of the results of an extensive geochronological study of an area in the Swiss Alps. The ages of biotites from metamorphic rocks were found using both the Rb–Sr and K–Ar methods. The rocks examined were originally formed or metamorphosed in the Palaeozoic and were finally involved in the Tertiary Alpine metamorphism. The results of this investigation are shown in Table 7.1. The two clearly older ages are evidently relict ages indicating the early history of the area. The sixteen remaining biotite samples, however, fall in age between 11 m.y. and 29 m.y. Such a range is far outside possible experimental error and must have definite physical significance. In fact Armstrong *et al.* observed that there is a geographical pattern to the ages

TABLE 7.1. K–Ar AGES OF ALPINE BIOTITES
(After Armstrong, Jäger and Eberhardt)

Locality	Age (m.y.)
Zervreila	335, 345
Monte Rosa	105
Truzzo	29·0
Bergell	24·0
Beura	21·0, 20·0
Gordemo	17·5, 17·5
Croppo	17·0
Soazza	17·5, 17·0
Claro	16·0
Acquacalda	15·5
Brione	16·0, 16·5, 16·0
Fibbia	16·5, 16·0
Cocco	21·5, 20·5
Riveo	13·0
Verampio	12·5, 12·5
Aeginental	14·0
Gletsch	16·0
"Gantergneiss"	14·0

found and concluded that the data result from a prolonged cooling history in this part of the Alps. From Rb–Sr ages on muscovites they considered that the Alpine metamorphism ended at least 30 m.y. ago and that the younger biotite ages and their spread result from the subsequent slow and steady uplift and consequent cooling of these minerals to temperatures at which they retain argon perfectly. Muscovite apparently becomes a closed system for argon and strontium diffusion at a higher temperature than does biotite. These authors noticed a "rather constant age difference of 8 m.y." between Rb–Sr ages of biotites and muscovites, the latter always being older. Harper (1964) in a study of the Scottish Caledonides also commented on a regular K–Ar age difference with muscovite always being older than biotite, in that instance by approximately 12 m.y.

Since the age resolution of the various methods is greater the younger the rocks, it seems that for rocks less than 500 m.y. in age, the question of the magnitude of the effect of deep burial and slow cooling on measured ages can be resolved by detailed studies. Whether or not cooling histories for terrains older than $1\cdot0$ b.y. will in many cases be lost in experimental uncertainty remains to be seen.

7.3. Diffusion Considerations

Having seen some of the geochronological observational evidence for argon and strontium loss by diffusion, we can now give briefly a few greatly simplified numerical considerations. We suppose that a spherical crystal begins its history with no argon content and some potassium which is uniformly distributed throughout. If argon is lost by continuous diffusion governed by Fick's Law, and if the concentration of argon is maintained at zero at the surface of the crystal, then we can write for the average Ar^{40}/K^{40} ratio in the crystal (Wasserburg, 1954)

$$\frac{Ar^{40}}{K^{40}} = \frac{6\lambda_e}{\pi^2} \sum_{n=1}^{\infty} \frac{1 - \exp - (n^2 a - \lambda)t}{n^2(an^2 - \lambda)} \qquad (7.1)$$

where

$$a = \pi^2 \frac{D}{a^2}.$$

D is the diffusion coefficient of argon in that mineral and is assumed to be a constant. The meaning of equation (7.1) may be readily grasped from Fig. 7.1, which shows a plot of Ar^{40}/K^{40} in the crystal versus time. Curves representing the result of continuous diffusion of argon lie to the left of the curve labelled $D = 0$ which represents no diffusion. For values of $D/a^2 > 10^{-21}$ sec^{-1} diffusional losses over a mineral's lifetime become measureable relative to the $\pm 5\%$ experimental error; D/a^2 values less than this have a negligible effect on a mineral's age. Hence, if definitive measurements on the diffusion coefficient for various minerals were available it would be relatively easy to make an approximate assessment of the effect of diffusional losses. Unfortunately the D values characterizing minerals are so small that laboratory measurements of them are very difficult to perform. It is customary to measure the D corresponding to higher temperatures. Low temperature estimates are then found by a straight line extrapolation over many orders of magnitude on a plot of log D versus $1/T$. The assumption is made here that in general we have $D = D_0\, e^{-E/RT}$ (where D_0 = a constant characteristic of the mineral, E = the activation energy for argon diffusion and R = the gas constant).

This is obviously a dubious process, however, since various factors affect the diffusion process and it is not by any means certain that such a straight line extrapolation is sound. It is likely, in fact, that the curve

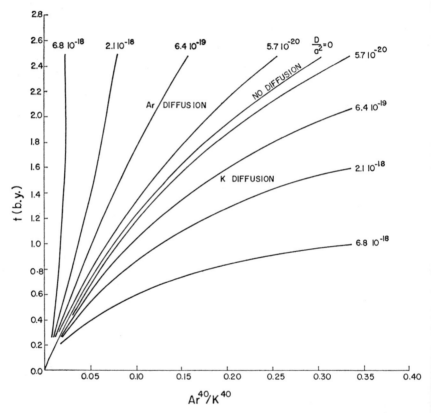

Fig. 7.1. Effect of continuous diffusional loss of Ar^{40} or K^{40} on Ar^{40}/K^{40} ratio as a function of time. D/a^2 in sec.$^{-1}$.

should generally consist of several segments of straight lines which differ in slope.

In Fig. 7.2 we show the data of Baadsgaard, Lipson and Folinsbee (1961) for sanidine from a bentonite. Straight line extrapolation indicates that at $25°C$ $D = 10^{-38}$ cm^2 sec^{-1}. Such a mineral would have $D/a^2 = 10^{-36}$ sec^{-1} if its radius were 1 mm. From our previous considerations, therefore, diffusional loss at this temperature over the mineral's lifetime would be negligibly small. On the other hand, these same investigators found for microcline at 25°C $D = 10^{-20}$ cm^2 sec^{-1}, if a similar speculative extrapolation were adopted. Reynolds (1957) estimated a minimum D for orthoclase-microcline of 10^{-19} cm^2 sec^{-1}. For assumed spherical felspar crystals 1 mm in radius these two estimates correspond to $D/a^2 = 10^{-18}$ and 10^{-17} sec^{-1}, which according to the curves of Fig. 7.1 would produce massive argon losses over the lifetime of the crystals. While it is

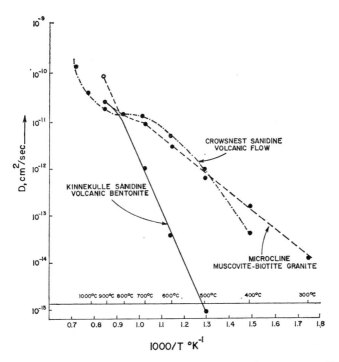

FIG. 7.2. Diffusion coefficients of felspars as a function of temperature. (After Baadsgaard et al., 1961.)

tempting to assign the felspar-mica K–Ar age discrepancy to this cause, it must be recalled that there seems to be no clear correlation between felspar age and percentage argon loss, whereas if steady diffusion were responsible the percentage daughter loss should increase with age. Another facet to the problem was revealed by Amirkhanoff et al. (1961), whose measurements indicated that argon in felspars is in several different "phases", each one characterized by its own activation energy. The temperature ranges during which these argon phases are released were found for one felspar to be: zero phase, $t < 400°C$; phase I, $400° < t < 700°C$; phase II, $600°C < t < 1100°C$. Amirkhanoff et al. proposed that the easily lost argon in felspar is contained in a surface zone and that if the argon and potassium in this zone were removed the remainder of the crystal would give reliable Ar^{40}/K^{40} ratios. Argon was removed from the surface zone by heating to 350–400°C for 2–3 hours while the surface potassium was removed by exchange with thallium. The striking results for four samples are shown in Table 7.2. In each case an originally discrepant felspar is brought up to concordance with the corresponding biotite. Further study of this effect is obviously merited.

TABLE 7.2. COMPARISON OF ORIGINAL K–Ar AGES OF
FELSPARS WITH TREATED FELSPARS AND INDEX MICAS
(After Amirkhanoff et al.)

Sample	Original felspar (m.y.)	Treated felspar (m.y.)	Associated mica (m.y.)
N1/1-5	1360	1575	1560
N823/5	1380	1860	1880
N6	160	275	285
N319	1720	2020	2010

Amirkhanoff et al. found also that the release of argon from micas was a complex process and noticed three argon phases in phlogopite. These phases became mobile in the temperature ranges 150–600°C, 750–900°C and > 900°C. Reynolds (1957) found evidence of two argon phases in lepidolite, one of which "is easily expelled by heating and is largely driven off at any temperature above 400°C, and another which is tightly bound and thus has a steep temperature dependence". Westcott (1966) found that the release of argon from biotite is associated with two activation energies. Below 800°C, the loss is best described by an activation energy $E \simeq 35$ kcal/mole, while above 800°C a value of $E \simeq 49$ kcal/mole is more appropriate.

It is evident that the mechanism of argon loss is only imperfectly understood and simple curves like those shown in Fig. 7.1 serve essentially to give the investigator a certain amount of physical insight into the orders of magnitude involved. The degree of uncertainty involved is perhaps well illustrated in Table 7.3 from Westcott (1966). The variation in estimates of the activation energy of argon diffusion in micas is huge and highlights the probabilities that diffusion is conditioned by slight variations in chemical composition, mechanical and thermal pre-history of the sample, and the technique of measurement. A detailed discussion of various methods of measuring diffusion coefficients is given by Fechtig and Kalbitzer (1966).

TABLE 7.3. ACTIVATION ENERGIES OF ARGON LOSS FROM MICAS

Mineral	E (kcal/g atom)	Method	Refs.
Muscovite	85	Laboratory	1
Phlogopite	67	Laboratory	1
Biotite	57	Laboratory	1
Margarite	54	Laboratory	2
Phlogopite	48	Laboratory	3
Muscovite	90	Laboratory	4
Biotite	35; 49	Laboratory	5
Biotite	27	Geological	6
Biotite	16	Geological	7
Biotite	< 28	Geological	8
Biotite	40	Geological	5

1. Gerling and Morozova (1957); 2. Fechtig *et al.* (1960, 1961); 3. Amirkhanoff *et al.* (1961); 4. Sadarov (1963); 5. Westcott (1966); 6. Hurley *et al.* (1962); 7. Tilton and Hart (1965); 8. Hart (1964).

Having seen the effect of diffusion over a mineral's full history and acknowledging the gross uncertainties in our estimates of diffusion coefficients it remains to make one interesting point in regard to the effect of the exponential variation of D with temperature. Suppose that over a certain temperature range we may write $D = D_0 e^{-E/RT}$ for a mineral. Then, if $E = 40$ kcal/mole, at 27°C (300°K) we have $D = 10^{-29} D_0$, while if $T = 127$°C (400°K) we have $D = 210^{-22} D_0$. Thus an increase in temperature of 100°C raises the diffusion coefficient by a factor of about 2×10^7, which means that the transition from a mineral being a wide open to a tightly closed system (and vice-versa) takes place over a very narrow temperature range. It is therefore meaningful to speak of a "critical blocking temperature" for a mineral below which temperature essentially all the argon is retained. (We have taken the term from the field of rock magnetism.) Thus the measurements of Harper (1964) indicate that the K–Ar "critical blocking temperature" of muscovite is higher than that of biotite. Evidently it required about 12 m.y. for the Scottish Caledonides to cool from the one blocking temperature to the other. The data of Armstrong *et al.* (1966) indicate that for Rb–Sr systems, the critical blocking temperature of muscovite is again higher than that of biotite. It appears that it required 8 m.y. for this part of the Alps to cool from the one blocking temperature to the other.

When an area is involved in a prolonged cooling period, the various radioactive clocks will commence recording at widely separated times as the temperature slowly falls from one critical blocking temperature to the next. Thus the Rb–Sr whole-rock and U–Pb zircon clocks will begin to record the lapse of time before any other systems. After this the other clocks will start in some sequence related to those shown in Table 5.2. A possible history such as this was adopted by McIntyre, York and Moorhouse (1967) to explain the distribution of radiometric ages found for the Madoc–Bancroft area of the Grenville province of the Canadian Shield. It was considered that while the major igneous and metamorphic events occurred 1100–1300 m.y. ago, because of the very slow cooling of the area the K–Ar biotite clocks did not begin to record time until 950–900 m.y. ago. This represents a much longer cooling interval than that adopted by Jäger (1965) and her colleagues for the Alps, but this presumably is a reflection of the different stages of erosion of the Grenville and the Alps. Clearly, as an orogenic belt cools, the radioactive clocks in the highest parts will begin to record time first as cooling will be most rapid near the surface. Cooling at great depths will obviously be slower. The Alps present rocks from a higher position in an orogenic belt than does the Grenville and consequently record much shorter cooling intervals. It is apparent that if critical blocking temperatures are as low as 250°C then long cooling histories and deep burial will have considerable importance, and mineral ages from complex terrains may be expected to postdate the peaks of metamorphism by many millions of years.

The recognition of the importance of the effects of long cooling histories on geochronological interpretations has spread widely in the 1960s. Apart from the references given earlier in the chapter, useful discussions of the problems have been given by Neuvonen (1961), Lambert (1964), Harper (1964, 1967) and Armstrong (1966b).

CHAPTER 8

THE PHANEROZOIC TIME-SCALE

8.1. Introduction

A relative time-scale for geological events was established in the last century by the use of stratigraphy and palaeontology. Local time-sequences were found by means of the stratigraphy of particular areas and these were correlated with a world-wide scale by the study of the fossils contained in the rocks. Such studies resulted in the subdivision of Phanerozoic time as shown in the first column of Fig. 8.1. Only with the development of the modern radiometric methods of dating has it been possible to affix actual times to such a scale with any degree of certainty. For an interesting review of the history of early attempts at quantifying the time-scale, reference should be made to Wager (1964). Reviews of more recent work are given by Holmes (1959), Faul (1960), Kulp (1961) and Evernden and Richards (1962). The most comprehensive analysis is contained in the volume *The Phanerozoic Time-Scale*, the proceedings of a symposium dedicated to Arthur Holmes for his outstanding contributions to geochronology. As this work appears as a supplement to the *Quarterly Journal of the Geological Society of London* it will subsequently be referred to as *Q.J.G.S.* (1964). The numerical time-scale suggested in the latter is shown in Fig. 8.1.

In principle the easiest way of assigning dates to a time-scale is to collect fossiliferous sediments characteristic of the opening and closing of the various periods and epochs and apply to these rocks the various modern radiometric methods. In practice this approach has not been followed to any extent because of the great difficulty experienced in obtaining reliable dates on sediments, which has already been emphasized in Chapter 5. Consequently two other approaches have been favoured. These are (a) the dating of volcanic rocks (including bentonites) which are interbedded with sediments of clear stratigraphic position; (b) the dating of igneous intrusives which cut sediments of known stratigraphic assignment and are themselves overlain by other stratigraphically well-assigned sediments.

(a) As has been discussed earlier (Chapter 4), volcanic rocks are dated in a number of ways depending on their composition. Usually minerals such as sanidine, biotite and plagioclase, and whole-rock basalts are examined with the K–Ar method. Occasionally the whole-rock Rb–Sr method has been used. An illustration of this approach is given by McDougall *et al.* (1966).

	m.y.			m.y.
CAINOZOIC			**PALAEOZOIC**	
QUATERNARY			PERMIAN	
Pleistocene	1·5–2 ??		Upper	240
			Lower	280
TERTIARY				
Pliocene	C. 7		CARBONIFEROUS	
Miocene	26		Upper	325
Oligocene	37–38		Lower	345 ??
Eocene	53–54			
Palaeocene	65		DEVONIAN	
			Upper	359
MESOZOIC			Middle	370
CRETACEOUS			Lower	395 ?
Upper	100			
Lower	136 ?		SILURIAN	430–440 ???
JURASSIC			ORDOVICIAN	
Upper	162		Upper	445 ?
Middle	172		Lower	C. 500 ?
Lower	190–195			
			CAMBRIAN	
TRIASSIC			Upper	515 ?
Upper	205		Middle	540 ?
Middle	215		Lower	570 ?
Lower	225 ?			

FIG. 8.1. The Phanerozoic Time-scale. (Modified after *Q.J.G.S.*, 1964.) Question marks indicate various degrees of uncertainty as discussed in the text.

(b) The dating of igneous intrusives is usually done by Rb–Sr whole-rock dating and the analysis of mineral separates by both Rb–Sr and K–Ar dating. A potential drawback to this method is that plutonic bodies may spend considerable time at depth, losing daughter isotopes all the time, thereby giving too young an age (Lambert, 1964). Apart from this restriction the method suffers from the fact that the intrusive is usually bracketed by sediments differing widely in age, so that the stratigraphic assignment of the body is usually much looser than the lavas of category (a). This method was used by Byström-Asklund *et al.* (1961). Harris *et al.*

(1965) presented an investigation where both (a) and (b) methods were applied.

8.2. Consideration of *Q.J.G.S.* Time-scale

In considering the reliability of the *Q.J.G.S.* time-scale several factors should be borne in mind. Approximately 380 ages were considered in the construction of the scale. Of these about 85% were K–Ar dates, 8% Rb–Sr dates and 4% U–Pb dates. Obviously it is desirable to have many more Rb–Sr and U–Pb figures. Over 85% of the K–Ar dates were less than 300 m.y. and only 20% of all dates were over 300 m.y. so it is not surprising that the older systems like the Cambrian, Ordovician and Silurian are the least well defined.

In his review of Cambrian data for *Q.J.G.S.* (1964) Cowie concluded that only fifteen ages merited detailed consideration and that "none is entirely satisfactory". His final best estimates were 495 m.y. for top, 515 m.y. for the Middle-Upper, 540 m.y. for the Lower-Middle and 570 m.y. for the base of the Cambrian system. Since those estimates were made, McCartney *et al.* (1966) have provided new data. These authors found a Rb–Sr whole-rock age of 574 ± 11 m.y. for the Holyrood granite in south-eastern Newfoundland. This granite intrudes the Harbour Main Group of volcanics and non-fossiliferous sediments which pass upwards into the Conception and Hodgewater–Cabot groups which pass upwards into fossiliferous Lower Cambrian strata. Since the Lower Cambrian rocks were not affected by the intrusion, it appears that the Holyrood granite is Late Precambrian on stratigraphic avidence. McCartney *et al.* therefore concluded that a tentative maximum age of 560 ± 11 m.y. was reasonable for the base of the Cambrian, noting that at least 15 m.y. would be required for the tectonic and sedimentary events occurring between the intrusion of the Holyrood granite and the deposition of the Lower Cambrian sediments. The Rb–Sr isochron age was calculated using the value $1 \cdot 47 \times 10^{-11} \, y^{-1}$ for the decay constant of Rb^{87}. If the equally popular value of $1 \cdot 39 \times 10^{-11} \, y^{-1}$ were used, and the same reasoning applied, presumably a tentative maximum age of the base of the Lower Cambrian would be about 590 m.y. Cowie's estimate remains consistent with this later work.

There is a similar paucity of Ordovician data. The only reliable dates considered in *Q.J.G.S.* (1964) were on bentonites from Kinnekulle (Sweden), Tennessee and Alabama representing Caradocian times (early Upper Ordovician), which indicated an age of about 445 m.y. There was no satisfactory means of fixing the upper and lower boundaries of the Ordovician. They were taken arbitrarily as about 430–440 m.y. and about 500 m.y. respectively. Harris *et al.* (1965) later presented K–Ar biotite data on the Bail Hill mica andesite and the Colmonell gabbro from the Scottish Ordovician and concluded that it was reasonable to assign minimum ages of 445 m.y. to the Lower Caradocian and 475 m.y. to the Arenig (Lower Ordovician). These data therefore give no grounds for changing the *Q.J.G.S.* (1964) estimates.

The Silurian has the worst radiometric coverage of all and Strachan's *Q.J.G.S.* (1964) summary consists entirely of eight dates on minerals from sedimentary rocks. Seven of these were on illites which were obviously anomalously young, and the last was on glauconite, which is a most unreliable age indicator. On recognizably most tenous grounds, therefore, the Silurian was considered to last from about 440 m.y. to 390 m.y.

Friend and House in *Q.J.G.S.* (1964) considered that the top of the Devonian was best dated at 345 m.y. This was based on the dating of the Chattanooga Shale, Tennessee, and the Snob's Creek Rhyodacite, Victoria, Australia. The Chattanooga Shale seems to be reasonably accurately located at the top of the Devonian, but the radiometric age picture is cloudy. Biotite from this formation gave a K–Ar age of 340 m.y. (Faul and Thomas, 1959) and a Rb–Sr age of 385 ± 40 m.y. (Adams *et al.*, 1958). U–Pb dating of the shale gave 350 m.y. (Cobb and Kulp, 1960). However, very large corrections for common lead were necessary so that no Pb^{207}/U^{235} age was calculated and the ages were based on the Pb^{206}/U^{238} ratios alone. One sample gave a Pb^{206}/U^{238} age of 450 ± 70 m.y. It seems clear that this shale merely provides evidence that 350 m.y. is a *minimum* age for the top of the Devonian. The Snob's Creek volcanics were considered to be 350 m.y. by Friend and House on the basis of K–Ar dating of biotites by Evernden and Richards (1962). A more intensive examination of these and other volcanics of this area by McDougall *et al.* (1966) confirmed values obtained by Evernden and Richards, but also revealed K–Ar biotite ages as high as 366 m.y. Rb–Sr whole-rock and felspar isochrons indicated ages of 367 ± 22 m.y. and 357 ± 10 m.y. using the $1 \cdot 47 \times 10^{-11} y^{-1}$ value for the decay constant of Rb^{87}. McDougall *et al.* therefore concluded that the evidence was very strong for revising the Devonian–Carboniferous boundary from 345 ± 10 m.y. to at least 362 ± 6 m.y. If the decay constant for Rb^{87} is taken to be $1 \cdot 39 \times 10^{-11} y^{-1}$, the indication would be that the top of the Devonian is at about 390 m.y. The same authors noted that the work of Cormier and Kelly (1964) on the Fisset Brook volcanics supported their contentions. On spore evidence the Fisset Creek Formation, Cape Breton Island, is earliest Mississipian (Lowermost Carboniferous), and Cormier and Kelly found a Rb–Sr whole-rock age of 358 m.y. if the $1 \cdot 47 \times 10^{-11}$ y^{-1} decay constant is used. This would become 380 m.y. using $\lambda = 1 \cdot 39 \times 10^{-11} \, y^{-1}$. Friend and House chose about 395 m.y. as the base of the Devonian. If this figure were retained, but the top of the Devonian were revised to 360–390 m.y., then only 5–35 m.y. would be left for the length of the Devonian system. Since about 50 m.y. has usually been considered necessary for the length of this system, then it would seem that the base of the Devonian may need revision to about 410–430 m.y. An examination of the Lower Devonian tie-points considered by Friend and House shows that such an extension is not impossible. Thus the Shap and Creetown granites yield Rb–Sr mineral ages of 420–425 m.y. if the $1 \cdot 39 \times 10^{-11} \, y^{-1}$ decay constant is used. While K–Ar biotite ages on presumed Lower Devonian granites and schists tend to cluster around 395 m.y., this may be a consequence of deep burial and prolonged daughter loss. The K–Ar biotite age of 393 m.y. (Haller and Kulp, 1962) for the Kap Franklin granite (Greenland) from the Middle Devonian is consistent with an extension of the base of the Devonian, as observed by McDougall *et al.* (1966). Such an extension would, of course, require adjustments to the Silurian and Ordovician boundaries. As we have seen, there is little evidence to preclude this.

Analysis of Francis and Woodland's data in *Q.J.G.S.* (1964) shows that a value of 362 ± 6 m.y. could readily be accommodated as the base of the Carboniferous. Currently there seems no reason to change the value of 280 m.y. adopted in *Q.J.G.S.* (1964) as the top of the Carboniferous. Webb and McDougall (1967), however, have suggested that the Permian–Triassic boundary be changed to 235 ± 5 m.y. from the *Q.J.G.S.* (1964) estimate by Smith of 225 m.y. Smith's estimate was based on data for New England intrusives in Australia whose upper stratigraphic limit was imprecise. Webb and McDougall found a K–Ar biotite age of 240 m.y. for a tuff with an apparently well-established Upper Permian stratigraphic assignment. Granites from the Maryborough Basin in south-eastern Queensland which cut Lower to Middle Triassic sediments gave concordant K–Ar and Rb–Sr whole rock ages of 218 m.y. when the value $1 \cdot 47 \times 10^{-11} \, y^{-1}$ was used for the decay constant of Rb^{87}. This implies that the Lower to Middle Triassic is older than about 220 m.y. These data combined with the tuff age to persuade Webb and McDougall to propose 235 ± 5 m.y. for the base of the Triassic. The date of the Palisade sill (New Jersey, U.S.A.) remains an important point for the Upper Triassic. It is apparently well located stratigraphically and was shown by Erickson and Kulp (1961) to have a probable K–Ar age of 193 m.y. Accordingly Tozer in *Q.J.G.S.* (1964) took 190–200 m.y. as being the best estimate for the Triassic–Jurassic transition. The Guichon Creek batholith was used by Holmes (1959) and Kulp (1961) to fix the Triassic–Jurassic boundary. This intrusion was assumed to have a K–Ar biotite age of about 180 m.y. However, White *et al.* (1967) have presented strong evidence to show that a better estimate of the K–Ar age is 200 ± 5 m.y. They also showed that the stratigraphic assignment of this batholith is imprecise and the intrusion could have been emplaced at any time between the Upper Triassic and the Middle Jurassic. It therefore has little weight in the calculation of the Triassic–Jurassic transition. Tozer's estimate of 190–200 m.y. for the latter remains reasonable.

Unequivocal dates from the Upper Jurassic are rare. The Shasta Bally intrusion (California), once considered a key sample, is now argued by Casey in *Q.J.G.S.* (1964) to be too imprecise in stratigraphic assignment to be definitive. The remaining Upper Jurassic dates are almost all on glauconites whose reliability as age indicators is questionable. The Jurassic–Cretaceous transition is therefore weakly defined, since the only Lower Cretaceous biotite ages (K–Ar) considered reliable by Casey in *Q.J.G.S.* (1964) are on intrusives in the U.S.S.R. whose significance is hard to assess from this point of view. The boundary was taken in *Q.J.G.S.* (1964) to be 136 m.y.

The Cretaceous–Tertiary boundary is somewhat better defined at about 65 m.y. by dates on Canadian and American volcanics (Funnell, *Q.J.G.S.*, 1964). Remaining much more obscure is the Tertiary–Quaternary transition (Pliocene–Pleistocene boundary). Funnell in *Q.J.G.S.* (1964) concluded that it would probable be ultimately found to lie between 1·5 m.y. and 3·5 m.y. Several estimates by later workers have in fact fallen in this range (Evernden and Curtis, 1965; Ericson *et al.*, 1964; Opdyke *et al.*, 1966; Curry, 1966; McDougall and Wensink, 1966; Stipp *et al.*, 1967). Using a combination of geological, K–Ar and palaeomagnetic methods Stipp *et al.* (1967) concluded that the Pliocene–Pleistocene boundary in New Zealand should be placed at about 2.5 m.y. This transition is, however, particularly obscure because of the use of two definitions, one based on biological change, the other on climatic alteration (Flint, 1965).

In summary it may be said that the numerical Phanerozoic Time-scale remains surprisingly imprecise. Definitive points are few and far between. The two necessary requirements of accurate stratigraphic assignment and reliable radiometric age give the impression of being virtually mutually exclusive, almost providing the basis for a geological uncertainty principle. It is apparent that many more dates on well-placed volcanics are needed and that the Rb–Sr whole-rock method of dating sediments such as shales may prove extremely important (Bofinger and Compston, 1967).

CHAPTER 9

THE PULSE OF THE EARTH

9.1. Introduction

It has been apparent for some time that the radiometric dates obtained on rocks from a single continent tend to cluster into definite groups. Ages are not uniformly distributed in time. Furthermore, the timing of the groups seems to be very similar from continent to continent, leading naturally to the supposition that the earth's history has been characterized by recurrent revolutions which have affected many parts of the world roughly simultaneously, when viewed on the appropriate time-scale.

In 1952 Holmes commented on the similar geochronological pictures presented by Australia, Africa and India. Since then the number of radiometric ages available for study has increased in almost an exponential fashion and, throughout this growth, the interpretations have remained strikingly constant. Thus certain numbers have taken on the appearance of "Magic Numbers". The apparent existence of these has been evident to many people in geochronology and it is not possible to assign their recognition to any one individual. Thus if we simply average the estimates of significant dates in the earth's history made by Hurley (1958), Voitkevich (1958), Davis and Tilton (1959), Gastil (1960), Goldich *et al.* (1961), Vinogradov and Tugarinov (1961), Sutton (1963), Fitch and Miller (1965) and Dearnley (1965), we get 2·69 b.y., 1·84 b.y., 1·09 b.y., and 0·57 b.y. The scatter of the estimates by individuals about these values is extremely small. Dearnley's estimates are based on by far the largest number of dates and provide the most up-to-date picture.

In Fig. 9.1 we reproduce Dearnley's plot of frequency of occurrence of a date versus age. Approximately 3400 age determinations are incorporated. The cumulative frequency curve shows abrupt changes in slope at the values 2·75 b.y., 1·95 b.y., 1·075 b.y., 0·65 b.y., and 0·18 b.y., and *on the simplest interpretation* would mark the onset of igneous and metamorphic processes in the earth. Apparently at each of these times the

116

earth became active, moved quickly to a peak of igneous and metamorphic activities and then gradually settled down until the next "Magic Number" signalled the onset of a new hiatus. There can be little doubt that the explanation of the significance of this group of "Magic Numbers" is of prime importance in earth sciences.

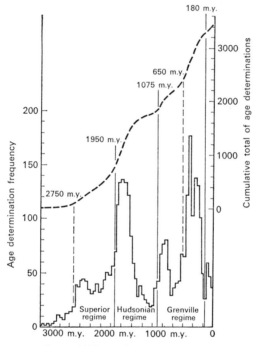

FIG. 9.1. Histogram of over 3000 age determinations. (After Dearnley, 1965.)

9.2. Runcorn's Hypothesis

Runcorn (1962) proposed a general hypothesis of the earth's geophysical evolution which provided a possible explanation of the Magic Numbers. The hypothesis involves an amalgamation of the ideas of Vening Meinesz, Chandrasekhar and Urey. Vening Meinesz (1952) concluded from Prey's (1922) spherical harmonic analysis of the earth's present topography that the positions of the continents reflect a regular pattern of convection in the mantle, the continents tending to station themselves over descending currents. The calculations of Chandrasekhar (1953) showed that the pattern of convection to be expected in the earth's mantle is determined by the ratio η of the radius of the earth's core to the radius of the earth. Since Urey (1952) had proposed that the earth began as a cold undifferentiated mass and has since been slowly growing a core, Runcorn pointed out that a gradually changing η would therefore cause changes in the patterns of mantle convection. As the core grew from zero radius to its present size, the various convection harmonics $n = 1$ up to $n = 5$ would be successively excited. Runcorn considered that at the time of transition from one mode of convection to the next higher, the continents would be under great stress and these transitions at critical values should be manifest in the geological record. He, in fact, associated the transitions between convection modes with the peaks in the distribution of age determinations as estimated by Gastil, i.e. the transitions were supposed to have the ages $2 \cdot 6 \pm 0 \cdot 1$ b.y., $1 \cdot 8 \pm 0 \cdot 1$ b.y., $1 \cdot 0 \pm 0 \cdot 1$ b.y. and $0 \cdot 2 \pm 0 \cdot 1$ b.y. Figure 9.2 illustrates the sequence of events contemplated.

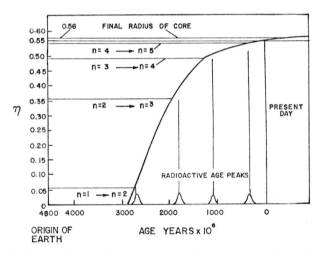

FIG. 9.2. Growth of the earth's core and change in mantle convection pattern with time. (After Runcorn, 1962.)

Whether this imaginative hypothesis is correct remains to be decided in the light of many more and varied data. While the $n = 4$ to $n = 5$ transition approximately 200 m.y. ago would correspond with Wegener's proposed onset of Continental Drift, there seems to be no convection transition to be associated with the Magic Number 0·65 b.y. Sutton (1963) has commented that Runcorn's hypothesis does "not appear to account for the distribution of orogenic belts, especially for the distribution in space and time of those formed over the past thousand million years".

9.3. Sutton's Chelogenic Cycles

Since the time of Hutton it has been realized that the earth has experienced a dynamic history and has undergone a succession of cycles of orogeny. Any one of these orogenic cycles may be described in a very generalized way by the following sequence of events:

(a) prolonged downwarping of a long belt of the earth's surface with the accumulation of sediments and volcanic rocks in the trough;

(b) metamorphism of the more deeply buried rocks in the geosyncline, involving the basement, sedimentary and volcanic rocks. Intense folding may occur accompanied by the intrusion of granitic rocks and volcanic activity;

(c) widening of the original geosyncline with repetition of (a) and (b);

(d) general uplift of much of the belt during epeirogeny. Erosion of these uplifted masses yields the conventional mountain system.

It appears that when any continent is sufficiently well examined geologically it is possible to distinguish the occurrence of at least ten such cycles (Holmes, 1965) and during the past 3000 m.y. there may have been more than fifteen orogenic cycles. The simple division of 15 into 3000 m.y. indicates that orogenic cycles have occurred, very roughly, every 200 m.y. A figure of this order of magnitude has in fact often been quoted as the interval between orogenies. Thus de Sitter (1959) said "We have seen that we know of orogenic periods occurring roughly each 200 m.y. since the Archean separated by periods of quiescence". Vening Meinesz (1951) spoke of particularly strong tectonic activity occurring at intervals of some 150–250 m.y. and considered these active episodes to be caused by convection currents involving the entire depth of the mantle. Poetic expression was given to the concept by Umbgrove (1947) with the words "What creature is this that breathes so heavily every 250 m.y.?"

If we now return to the Magic Numbers 2·75, 1·95, 1·08, 0·65, 0·18 b.y. (Dearnley's values), we see that they are separated from each other by the time intervals 0·80, 0·87, 0·43, 0·47 b.y., which are far greater intervals of time than the 200 m.y. usually considered to elapse between orogenic cycles. If these larger intervals are not spurious, then they indicate a rhythm in the earth's history which is much slower than that associated with the recurrence of the orogenic cycles. To describe this longer rhythm Sutton (1963) coined the term "Chelogenic Cycle".

A Magic Number marks the onset of a chelogenic cycle which runs to completion by the time the next Magic Number appears signalling the next chelogenic cycle. In Sutton's opinion the earth has undergone three successive long-term cycles in the past 2·8 b.y. These are:

(a) the Shamvaian cycle, 2·8–1·8 b.y.;
(b) the Svecofennid cycle, 1·8–1·1 b.y.;
(c) the Grenville cycle, 1·1 b.y.–present.

Clearly any chelogenic cycle must contain approximately four or five orogenic cycles. Sutton stressed that the main structural geological units throughout the world apparently commenced their development with the onset of one of these long-term cycles. This argument is illustrated by Table 9.1, which shows the range of dates for each structural province

TABLE 9.1. RANGES OF DATES IN TECTONIC PROVINCES
(After Sutton)

Locality	m.y.	m.y.	m.y.
Scotland	2800–2200	1700–1300	1000–750 –400
Baltic	2900–1900	1900–1250	1100–400
Central Africa	2700–2000	1850–1500	1100–475
Canada	2750–2300	1850–1550	1200–800
Western U.S.A.	2700–2350	1750–1200	1000– 18
East Asia (N. China; Aldan)	2700–2250	1900–1100	1100–650
India	2450–2300	1650–1300	1150–735
Australia	2950–2300	1700–1000	1100–100
Antarctica		1800–	–375
Central N. America	2800–2000	2000–1500 –1200	1200–500

of the major Precambrian terrains of the world. While recognizing that any geological province usually bears the evidence of having undergone several orogenic cycles, Sutton notes that these orogenic belts rarely truncate each other and "never abruptly cut off an earlier belt of the same cycle. The various fold-belts of a single (chelogenic) cycle are usually found in close proximity in several of the continents, and it is this spatial grouping of fold-belts formed over a certain time-span which gives rise to the great structural provinces such as those of the Canadian Shield." Sutton called this unit, formed by grouping orogenic cycles, a chelogenic cycle since this may be translated as a "shield-producing" cycle.

In their earlier paper Vinogradov and Tugarinov (1961) discussed the Baltic and Ukrainian shields and the Russian Platform in great detail and clearly mentioned long-term cycles. To distinguish these from orogenic cycles they suggested the term "megacycle" and commented, "The formation of both shields took place during a long period of geological time,

TABLE 9.2. TYPICAL CHELOGENIC CYCLE

(After Sutton)

Stage	State of continents	State of mountain chains	Possible state of convection system in mantle
4	Two clusters develop once more	Further restriction of orogeny	Development of new world-wide cell from major cells of stage 3
3	Maximum dispersion of continental masses	Further restriction of orogeny	Major cell under each fragmenting continent. No single cell of world-wide extent
2	Fragmentation of the clusters starts	Restricted to outer parts of continent	Major cell under each continent. World-wide cell weak or non-existent
1	In two clusters	Network over continent. Orogeny returns to previously non-orogenic regions	Many cells of sub-continental size below each continent + world-wide cell originated in previous cycle
Start			

(left margin label: 750–1250 million years)

from 3,500 to 1,900 m.y. ago. The most intensive periods of magmatism were 3,200 ± 300, 2,600 ± 200 and 1,900 ± 100 m.y. This confirms the existence of great megacycles in the geological history of the earth's crust with the period of renewal every 600 m.y." The proposed subdivision of geologic time by Vinogradov and Tugarinov and Sutton may be seen in Table 9.3.

Sutton proposed the structure for the long-term cycle which is shown in Table 9.2. The detailed continental drift and convection current aspects are obviously speculative, and the most interesting aspect of Table 9.2 is the emphasis on the rapid onset of orogeny followed by the gradual withdrawal of this activity to the peripheral regions of continents.

TABLE 9.3. DIVISION OF GEOLOGICAL TIME INTO MEGACYCLES OR CHELOGENIC CYCLES

m.y. ago	Vinogradov and Tugarinov	Sutton	Fitch and Miller
	Phanerozoic	Grenville	Alpine
			Varisco-Caledonian
1000	Upper Proterozoic		
	Lower Proterozoic	Svecofennid	Grenville
2000			
	Archaean	Shamvaian	Churchill
3000			
	Katarchaean		Superior
4000		?	
	?		Pre-Superior
4550			

9.4. Gastil, Fitch and Miller and Anorogeny

In 1960 Gastil published a compilation and an analysis of 413 radiometric ages and his well-reasoned discussion has had considerable influence on the later writers in the West. Gastil stressed that mineral ages were distributed in groups rather than uniformly through time. The average length of the groups was 240 m.y. with a standard deviation of 60 m.y. These groups were largely independent of the age-method used and agreed from continent to continent. The intervals between peaks of mineral date abundances showed a surprisingly small degree of scatter and the average interval between peaks was 417 ± 47 m.y. (s.d.). Gastil was therefore led to compare the age distribution with a 420 m.y. cycle in which 210 m.y. intervals of mineral date abundance alternated with like intervals of mineral date scarcity. He concluded that if cycles are indicated by the distribution of ages then they represented the "cyclic withdrawal of large portions of the earth's crust from active igneous and metamorphic mineralogenesis".

Fitch and Miller (1965), in reviewing the evidence for the existence of megacycles, placed strong emphasis on this aspect of Gastil's arguments. Furthermore, in the five years which had elapsed since Gastil's publication, the methods of geochronology had been developed to the point where they could be applied to basic rocks. Thus there were newly available a number of dates on some of the many dyke swarms of the world. Fitch and Miller observed that these tended to fall into the gaps of the earlier age distributions which were essentially based on dates from orogenic environments. This led them to speculate that the state of the earth's crust during a megacycle changes from one of tension during the anorogenic phase to one of compression during the orogenic phase. In 1925 Joly had reached similar conclusions, but in his case it was in regard to the shorter-term orogenic cycles. The division of time into megacycles proposed by Fitch and Miller is shown in Table 9.3. Since they chose to make the anorogenic phase the first part of a megacycle their proposed arrangement is out of phase with the others shown.

The approach of Fitch and Miller (1965) has much to commend it with its emphasis on separating dates obtained on "orogenic bodies" from those characteristic of anorogenic environments. As we mentioned earlier, until recently the vast majority of published ages referred to rocks formed in an orogenic surrounding because of the ease of analysis of minerals found in such bodies. The many dykes, flood basalts and alkaline complexes have only recently been subjected to detailed age analysis. Yet these latter have a distinct potential geochronological advantage over rocks such as gneisses, schists and synorogenic granites which are characteristically formed in the orogen. For, as discussed in Chapter 7, the various minerals formed in the heart of the deeply buried, active orogen may not be recording the times of crystallization and the various folding phases, so much as the time of final uplift and cooling. And obviously this could cause a considerable smearing of the age distributions. In contrast, the rocks formed in an anorogenic environment are usually emplaced into relatively cool crustal rocks in a stable shield area and cool rapidly below critical blocking temperatures. In the absence of subsequent metamorphisms then, such rocks should be the ones most likely to record times of crystallization. Somewhat in this vein, Fahrig and Wanless (1963) in a K–Ar study of diabase dykes from the Canadian Shield said:

> We suggest that intense folding and compression in the orogenic belt and less intense compression within the craton results from sub-crustal currents. If relaxation occurs during or after a major orogenic cycle, as a result of diminution in size of sub-crustal currents, diabase dyke intrusion may occur both in the orogenic zone and in the adjacent stable craton. If a period of tension is succeeded by additional mountain building accompanied by folding, metamorphism and granitic intrusion in the orogenic belt, diabase dykes relating to the preceding period of tension will be preserved or recognizable only in the stable craton. A series of dyke swarms in the craton may reflect more intense rhythmic compression and relaxation in neighbouring developing mobile belts.

This approach is not without its own drawbacks, however. Such dykes are usually dated by K–Ar analysis of samples of the whole rock since separation of suitable minerals is hampered by the fine grain. That this method may yield low ages is always a possibility on account of the chance of argon leakage from the felspar. According to Fahrig and Wanless (1963), the best material for whole-rock K–Ar dating of dykes is to be found in the chilled contact zone, and this observation is borne out by the recent work of Leech (1966).

However, despite the definite problems involved in dating fine-grained basic rocks, the concept of elucidating what went on inside an active orogenic zone by examining the igneous activity in the adjacent relatively stable cratons is a most important one. It may well be that the alkaline intrusives offer useful prospects in this regard. Thus numerous small alkaline complexes are found in the Superior province of the Canadian Shield. While the country rocks, consisting of gneisses and granites, are generally characterized by ages of approximately 2·5 b.y., the alkaline rocks fall into at least two groups at approximately 1·07 b.y. and 1·7 b.y., as was observed by Gittins, Macintyre and York (1967). Since these dates are similar to those obtained on rocks from the Grenville and Churchill provinces respectively, it seems likely that these intrusions into the adjoining stable Superior craton reflected some aspects of these later orogenic activities.

Much work remains to be done on non-orogenic rocks before their correct relation to orogenic activity is found. Nonetheless, because of their rapid cooling at formation and their often subsequently peaceful history in a stable craton they may well prove much more amenable than are orogenic rocks to world-wide correlation.

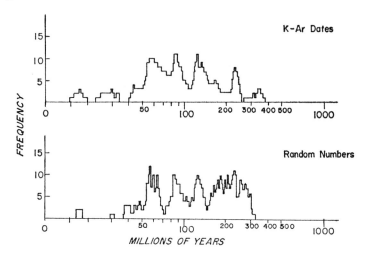

FIG. 9.3. Comparison of histograms of K–Ar ages with a set of random numbers (After Ross, 1966.)

9.5. Cautionary Tail

We have seen that there is a growing tendency to group orogenic cycles into clusters and make them components of a longer-term rhythm. It behoves us now to add some reservations.

At the outset it should be acknowledged that lumping large numbers of ages into one histogram, and attaching major geological significance to any ensuing structure, is a risky business albeit exciting. The effect of long cooling histories on such histograms, for instance, remains to be seen (see Chapter 7). Ross (1966) presented a useful reminder of the dangers which we illustrate in Fig. 9.3. In the upper portion of the diagram is a histogram plot of 129 K–Ar dates taken from a paper by Gabrielse and Reesor (1964) on the geochronology of the Canadian Cordillera. Beneath this are plotted the first 129 random numbers, having values 350 or less, from a table of random numbers having a uniform distribution (Kendall and Smith, 1939). There is obviously a striking resemblance between the two histograms and it is undeniable that a useful point is made. However, the persistence of the Magic Numbers, defined earlier, as many more data have been added, and the similarity of the geochronological pictures of the different continents indicate that a genuine structure is to be seen in the age distributions. Undoubtedly the presently existing sample and geographical bias must cause some distortion but this effect should be minimized in the future.

Further interesting discussions regarding the timing of orogenies and the building of continents may be found in Gilluly (1949), Wilson (1951), Wasserburg (1961, 1966), Hurley *et al.* (1962) and Engel (1963).

CHAPTER 10

REVERSALS OF THE EARTH'S
MAGNETIC FIELD

10.1. Introduction

The earth's magnetic field would be reproduced to a good approximation by a dipole located at the centre of the earth, having its axis tilted about 11° from the rotation axis.

The major component of the field is produced within the earth, as was first shown by Gauss by spherical harmonic analysis. Since typical Curie temperatures would be easily exceeded at moderate depths it is reasonable to associate the field's existence with the core, the most likely theory of the origin of the field being due to Elsasser (1939) and Bullard (1949) who proposed that complex motions in the highly conducting liquid outer core are capable of sustaining a field of the type observed. Unfortunately the intricacies of the problem have defied rigorous analysis and there still remains considerable mystery even though the magneto-hydro-dynamic model is now quite well accepted.

Recently, features of the behaviour of the geomagnetic field have been emerging which are of considerable interest and with which the dynamo theory is consistent. Palaeomagnetic measurements have shown that many rocks a few million years in age have been magnetized in a direction almost exactly 180° away from the direction of the present earth's field. These are described as "reversely magnetized" rocks. A considerable number of examples exist of volcanic sequences of a variety of ages in which there are two principal directions of magnetization 180° apart. Two simple explanations immediately occur;

(a) the earth's magnetic field actually reverses in direction by 180° every so often and rocks acquire "normal" or "reversed" magnetization depending on the field direction at the time of acquisition of their magnetization; or

(b) the earth's field has not reversed episodically, but a number of rocks can acquire magnetization in a direction which is 180° away from the ambient field.

Reversed magnetization acquired as in (b) is called a "self-reversal" and in fact this phenomenon is now known to occur in nature. In the case of the Haruna dacite, examined by Nagata and Uyeda (Nagata, 1961), the rock is reproducibly self-reversing in the laboratory. Despite this, less than 1 % of all rocks tested in the laboratory have been shown capable of self-reversal. While such an observation might make hypothesis (b) seem highly unlikely, it must be noted that Uyeda and others have shown that self-reversal can be critically dependent on chemical composition and rates of cooling of the rocks being examined. Hence, since laboratory experiments must inevitably be carried out on an extremely short time-scale their results must be interpreted with some reserve.

Probably the best way of deciding between field reversal and self-reversal is to adopt a combined palaeomagnetic and geochronometric approach. If the earth's magnetic field were reversed during some interval, rocks, regardless of their locality, would be formed during this time with reversed magnetizations. During intervals when the field was normal, rocks would form with normal directions of magnetization. In this way rocks of a given age should be either all normally magnetized or reversely magnetized irrespective of their place of formation in the world, assuming for the moment that the reversal process is instantaneous. Detailed studies of young rocks by Bruckshaw and Robertson (1949), Roche (1953), Hospers (1953, 1954) and Campbell and Runcorn (1956) showed that roughly half the rocks examined were normally magnetized and half were reversed. Roche and Hospers were led to believe that the earth's field reversed several times in the Tertiary, the reversal process occupying about 10,000 years.

10.2. K–Ar and Palaeomagnetic Studies

Recently Doell *et al.* (1966), McDougall and Chamalaun (1966) and Cox (1969) have reviewed the latest evidence. This comprised essentially the results of their own K–Ar and magnetic measurements on volcanic rocks carried out since 1963. Very suggestive and extremely interesting results were obtained and they are illustrated in Fig. 10.1. Rocks in the age ranges (i) 0–0·7 m.y. were mostly normally magnetized, (ii) 0·7–2·4 m.y. rocks were predominantly reversely magnetized, (iii) 2·4–3·3 m.y. rocks were mainly normally magnetized, (iv) 3·3–4·5 m.y. rocks were mainly reversely magnetized. These relatively long periods have been called "epochs" (Cox *et al.*, 1964) and have been named (i) Brunhes, (ii)

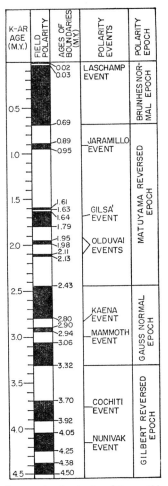

FIG. 10.1. Time scale of reversals of polarity of the earth's magnetic field as determined by K–Ar dating. (After Cox, 1969.)

Matuyama, (iii) Gauss and (iv) Gilbert after scientists who have made significant contributions to our knowledge of terrestrial magnetism. It may be seen in Fig. 10.1 that several much shorter periods of constant geomagnetic polarity have also been found. Cox *et al.* (1964) have named such shorter periods "events". Seven of these have so far been discovered: (i) the Jaramillo normal event lasting from approximately 0·95 to 0·89 m.y. ago; (ii) the Gilsa normal event approximately 1·61–1·79 m.y. ago; (iii) the Olduvai normal event approximately 2·0 m.y. ago; (iv) the Kaena reversed event about 2·85 m.y. ago; (v) the Mammoth reversed event about 3·0 m.y. ago; (vi) the Cochiti normal event about 3·8 m.y. ago; (vii) the Nunivak normal event about 4·1 m.y. ago. (See Cox, 1969.) There is also some evidence (Bonhommet and Babkine, 1967) of a reversed event, the Laschamp, ending about 20,000 y ago (Bonhommet and Zähringer, 1969), and another reversed event, the Blake, 108,000–114,000 y ago (Smith and Foster, 1969).

The evolution of such a clear-cut time-scale of reversals of polarity of the geomagnetic field is a striking substantiation of the suggestions of field reversal proposed by the early palaeomagnetic investigators Brunhes (1906), Chevallier (1925), Mercanton (1926) and Matuyama (1929), and more recently by Roche and Hospers.

The above palaeomagnetic study was only possible because of the refinement of the K–Ar method of age determination to the level where it could handle basic volcanic rocks, low in potassium, less than 10 million years in age and therefore containing extremely small amounts of argon. The non-random nature of the results suggests that the K–Ar figures for such rocks are reliable. It is not easy, however, to estimate the time occupied by the actual reversing process. Because very few rocks in the age range considered show directions other than normal or reversed it is evident that the reversal occurs rapidly, probably in the course of about 5000 y (Cox and Dalrymple, 1967). Stratigraphic arguments and studies of marine cores also indicate that of the order 10^4 y are required for the actual process of reversal to take place. Differences of this order are, unfortunately, still beyond the resolution of the K–Ar age measurements.

As efforts are made to extend this field reversal time-scale further into the past, more obstacles will be met (Baksi *et al.*, 1967; Cox and Dalrymple, 1967). The accuracy of K–Ar analyses will remain at about ±5% for rocks up to 50 m.y. old. Thus when rocks 20 m.y. old are being examined the age uncertainty will be ±1·0 m.y. which, in recent times at least, seems to be about the length of a stable magnetic epoch. For rocks 50 m.y. old the age uncertainty is at least ±2·5 m.y., and a number of reversals could occur within this uncertainty time. Furthermore, both the magnetization and potassium and argon contents of the rocks are more likely to have been altered by geological disturbances as the age of the rocks studied increases. However, the present apparent time-scale for reversals may not be typical, for no reversely magnetized rocks of Upper Permian to Upper Carboniferous age have been found so far (Irving, 1964). Should this situation prevail, then between the approximate times 230 m.y. and 280 m.y. ago the earth's field apparently did not reverse itself. It seems possible that over tens of millions of years the field may remain approximately constant in direction and then may go into an unsteady state during which times reversals occur every one million years or so. The behaviour may be somewhat analogous with that of the ice ages, although on a different time-scale. In the past one million years the ice cap has reached down far into the lower latitudes four times, receding each time. However, from the Pleistocene back to the Permian no major ice advances are recorded. In Permian and Carboniferous times, however, numerous glaciations occurred. There was glacial quiescence going further back until Eo-Cambrian times when glacial episodes were widespread. Possible connections between tectonism, palaeoclimates, and frequency of field reversal were discussed by Hide (1967) and Irving (1967).

It may be said in conclusion that while self-reversal undoubtedly occurs naturally, these recent combined palaeomagnetic and geochronometric analyses strongly suggest that episodic field reversal has been a definite feature of the earth's history. Any theory proposed to explain the magnetic field of the earth must therefore provide a ready mechanism for frequent reversals of the field. The dynamo theory is apparently able to allow reversals and its claims are considerably strengthened as a result.

The pattern of field reversals described above, which was deduced from the combined application of K–Ar dating and palaeomagnetic methods to volcanic rocks, has received independent confirmation from studies of sea bed cores. Opdyke *et al.* (1966) found that the direction of magnetization of such cores alternated from normal to reversed along their length. Normal sections of a core were deposited when the earth's magnetic field had a normal direction, reversed sections were laid down during time of a reversed field. The pattern of normal and reversed magnetization along the cores corresponds well with that in Fig. 10.1, although it is somewhat distorted by the lack of constancy in rates of sedimentation (Fig. 10.2). There is also some debate about whether or not the Gilsa event is recorded in the cores.

10.3. Ocean Floor Spreading

The concept of reversals of geomagnetic field polarity has been applied with striking success to the theory of continental drift. Since 1958 it had been known that there were curious anomalies at sea in the geomagnetic

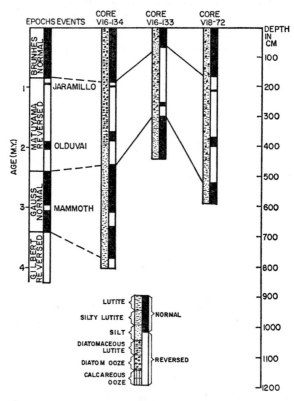

FIG. 10.2. Correlation of magnetic stratigraphy in three cores from the Antarctic with the magnetic field reversal time-scale. (After Opdyke *et al.*, 1966.)

field (Mason, 1958). These took the form of parallel strips where the field strength was alternately lower and higher than the average in the area. Such strips were found to be roughly parallel to and symmetrical about oceanic ridges. Vine and Matthews (1963) proposed that these anomalies were due to strips of alternately normally and reversely magnetized volcanics on the ocean floors. These were generated in accord with the ocean floor spreading theory of Holmes (1928), Hess (1962) and Dietz (1961). In this theory, convection currents in the earth's mantle bring material to the surface to be poured out at the sites of oceanic ridges where it cools to form new ocean floor. The ocean floor is meanwhile splitting along the ridges and the new material is ferried away on both sides from the ridges. When the volcanic material cools through the Curie temperatures of its constituent minerals it will acquire magnetization in the direction of the ambient geomagnetic field. Vine and Matthews pointed out that since the earth's field is alternately normal and reversed, the lavas being carried away from the ocean ridges will be alternately normally and reversely magnetized. The ocean floor will thus gradually be made up of stripes of

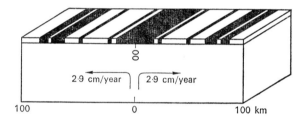

FIG. 10.3. The Vine–Matthews mechanism for generating oceanic magnetic anomalies.

alternately normally and reversely magnetized basalts. As the ocean floors are supposed to be spreading symmetrically away from the ridges, the pattern of striping should also be symmetrical about the ridges. The Vine–Matthews mechanism is illustrated in Fig. 10.3.

Vine and Wilson (1965), Vine (1966), Pitman and Heirtzler (1966) and others have matched the geomagnetic reversal time-scale of Cox, Doell, Dalrymple and McDougall with the oceanic anomalies patterns with considerable success, when ocean floor spreading rates of the order 1-10 cm/y are assumed. These are just the rates called for by numerous advocates of continental drift over the years (see particularly Holmes, 1965, p. 1032) and it is now considered by a vast majority of opinion that ocean floor spreading is essentially proved in its broad outline. Undoubtedly much revision and addition of detail are to be expected. The oceanic anomalies patterns indicate that the magnetic field reversal time-scale extends backwards in time well beyond what has so far been established

by K–Ar dating, such an extrapolated scale due to Heirtzler being shown
in Fig. 10.4. The testing of the details of such scales by K–Ar dating is
highly desirable, to shed light on the rate of ocean floor spreading, but
becomes very difficult for times greater than 10 m.y. ago because of the
lack of precision of the K–Ar method (Baksi *et al.*, 1967). It is obvious that
this topic will become an even more significant part of K–Ar dating than
it is now.

Fɪɢ. 10.4. Geomagnetic polarity reversal time-scale extrapolated backwards in
time with the aid of oceanic magnetic anomalies. (After Heirtzler.)

CHAPTER 11

METEORITES

11.1. Introduction

Material is continually arriving at the earth's surface from outer space. Estimates of the actual amounts vary from thousands to millions of tons per year. The individual pieces of matter reaching the earth's surface are termed meteorites and have long been studied by geologists. It was soon realized that these extra-terrestrial objects could be divided into three groups—stones, stony-irons and irons.

STONES

These consist in bulk of the silicate minerals, olivine and pyroxene, and are further subdivided into the categories *chondrites* and *achondrites*. Chondrites are characterized by the presence of small ($\lesssim 3$ mm), roughly spherical olivine (sometimes with pyroxene) aggregates. Achondrites do not possess these. The stones generally have small amounts of nickel–iron alloy.

STONY-IRONS

These possess approximately equal contents of silicates and nickel–iron alloy.

IRONS

These bodies are essentially a nickel–iron alloy with very small amounts of silicate minerals.

The meteorites have fiery passages through the earth's atmosphere and frequently break up into small fragments. Devastating impacts can result from the fall of the largest objects. The renowned Meteor Crater in north-central Arizona is 600 feet deep, is encircled by a rim of height varying between 100 and 200 feet above the plains and is approximately three-quarters of a mile in diameter. It has been estimated that the meteorite responsible struck the earth at 15 km/sec and weighed about 60,000 tons (Shoemaker, 1963). Fortunately such catastrophic arrivals are rare on the human time-scale.

No adequate theory regarding the origin of meteorites exists. They are often supposed to have come from the asteroidal belt in the solar system, the asteroids being thought of as fragments of a disrupted planet. The irons would be representative of the core of such a primeval body, the stones being remnants of the mantle and crust. However, the mass of material in the asteroidal belt is not enough to form a planet of reasonable size and it is also very difficult to postulate a satisfactory disruptive mechanism. Nor is the sand-blasted appearance of the chondrites easily accounted for (Mason, 1962; Wood, 1963; Anders 1964; Urey, 1964).

However, as we shall see, the dating of these various objects, while not without its own difficulties of interpretation, indicates that the various objects underwent solidification and chemical differentiation about 4·6 b.y. ago.

11.2. Dating Meteorites
K–Ar

Since the early determinations by Gerling and Pavlova (1951), many K–Ar ages have been reported for stone meteorites. Not all the ages in the literature are equally reliable, principally because of the difficulties in determining the very low potassium contents and the vigour with which the meteorites have to be heated to secure complete argon extraction. The ages show a spread from about 400 m.y. to about 5 b.y. There is, however, a significant clustering of ages between 4·0 and 4·8 b.y. which is clearly visible in Fig. 11.1.

The method is full of hazards in its application to the iron meteorites. Both Ar^{40} and K^{40} are produced in iron meteorites by cosmic ray bombardment. No reliable way of correcting for these effects is known. Furthermore, the concentrations of Ar^{40} and K^{40} are extremely low and present great experimental problems. The first K–Ar study on these bodies was reported by Stoenner and Zähringer (1958), who found "ages" ranging from 5 to 13 b.y. Their results have been substantiated by Fisher (1965), Muller and Zähringer (1966) and Rancitelli et al. (1967). It is difficult to reconcile these data with the Rb–Sr ages of 4·25–4·75 b.y. obtained by Burnett and Wasserburg (1967b) on iron meteorites, with the Pb–Pb dating of meteorites by Patterson (1956) and with the evidence of the trapping of

I^{129} in meteorites (Reynolds, 1960), and their interpretation must await the results of more experiments on meteorites.

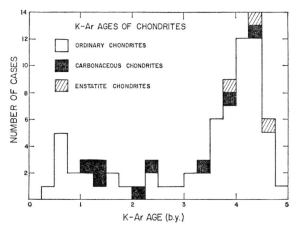

FIG. 11.1. Histogram of K–Ar ages of stone meteorites. (After Anders, 1963.)

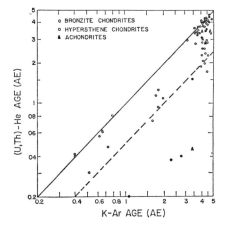

FIG. 11.2. U,Th–He ages versus K–Ar ages of stone meteorites. (After Anders, 1964.)

U–He, Th–He

He³ and He⁴ are both produced in meteorites as a result of cosmic ray bombardment and allowance has to be made for this He⁴ in the determination of U–He and Th–He ages. The correction for such cosmogenic

He^4 in chondrites usually amounts to less than 10% of the total He^4. In the achondrites the correction is sometimes greater than this, but few measurements are available. As with the K–Ar ages there is a considerable spread among the results, from about 400 m.y. to about 5 b.y. (Fig. 11.2), but there is a greater preponderance among the lower ages in the U–He, Th–He data. The occurrence of a number of U–He ages greater than 4 b.y. suggests that a re-examination of this method as applied to terrestrial rocks would be worth while.

There is some suggestion of a clustering of U,Th–He ages of hypersthene chondrites at about 0·5 b.y. (Anders, 1964). Support for this was found by Turner et al. (1966), who applied the Ar^{39}/Ar^{40} method with stepwise heating to the Bruderheim meteorite. Their results indicated that Bruderheim lost 90% of its radiogenic Ar^{40} 495 ± 30 m.y. ago.

The iron meteorites have extremely low uranium contents, 0·003–0·3 ppb (Reed and Turkevich, 1955; Reed et al., 1958) and their helium content is virtually entirely cosmogenic. No reliable dates therefore have been found with this method. It may be that the silicate nodules in some iron meteorites will prove suitable for U,Th–He dating just as these nodules enabled Burnett and Wasserburg (1967a, b) to date iron meteorites with the Rb–Sr method.

Rb–Sr

The Rb–Sr method was first applied to meteorites by Schumacher (1956), Herzog and Pinson (1956) and Webster et al. (1957). These and subsequent studies have combined to show that a major chemical differentiation occurred among these bodies 4·3 to 4·7 b.y. ago. Between 1960 and 1965 the usual approach was to measure the Sr^{87}/Sr^{86} and Rb^{87}/Sr^{86} ratios for a number of meteorites, then plot these on the Rb–Sr whole rock diagram and calculate the age of the meteorites from the slope of the isochron as discussed in Chapter 5. It was thus necessary to assume that all the meteorites (or at least two in the earlier calculations) at one time had a common Sr^{87}/Sr^{86} ratio. In this way Gast (1962) used four achondrites and five chondrites to find an age of about 4·5 b.y. and an initial ratio of Sr^{87}/Sr^{86} = 0·700. The achondrites have very low Rb/Sr ratios and their Sr^{87}/Sr^{86} ratios have therefore been changed only minutely in 4·5 b.y. These ratios were on average 0·701 when normalized by Gast to Sr^{86}/Sr^{88} = 0·1186, and corrected for radiogenic Sr^{87} addition in 4·7 b.y. The more conventional Sr^{86}/Sr^{88} = 0·1194 normalization gives the initial value for achondrites as $Sr^{87}/Sr^{86} = 0.699$. Gast concluded that the stone meteorites lay between 4·3 and 4·7 b.y. in age, using $\lambda = 1·39 \times 10^{-11}$ y^{-1} as the decay constant of Rb^{87}.

Compston *et al.* (1965) performed the first Rb–Sr age measurement using different phases of a single meteorite (the Bishopville aubrite). This approach considerably minimizes the importance of the assumption regarding common initial strontium isotope ratios. This method was used by Wasserburg *et al.* (1965) and Burnett and Wasserburg (1967a, b) for silicate nodules extracted from single iron meteorites. Their results for the Weekeroo Station iron meteorite are shown in Fig. 11.3. Bogard *et al.* (1967) were able to obtain a single meteorite isochron age for the Norton County achondrite, while Shima and Honda (1967) used a fractional dissolution method to find single meteorite ages for two chondrites.

TABLE 11.1. Rb–Sr AGES OF METEORITES

Chondrites (b.y.)	Achondrites (b.y.)	Irons (b.y.)	Author
4·8 ± 0·4			Schumacher (1956)
4·7			Herzog and Pinson (1956)
4·6 ± 0·44			Webster *et al.* (1957)
4·3 − 4·7			Gast (1962)
4·46 ± 0·35*			Murthy and Compston (1965)
	3·7 ± 0·2		Compston *et al.* (1965)
4·52 ± 0·12			Pinson *et al.* (1965)
4·45 ± 0·03			Shields *et al.* (1966)
		3·8 ± 0·1	Burnett and Wasserburg (1967a)
		4·25 − 4·75	Burnett and Wasserburg (1967b)
	4·7 ± 0·1		Bogard *et al.* (1967)
4·38 − 4·70			Shima and Honda (1967)

Ages calculated using $\lambda = 1·39 \times 10^{-11}$ y^{-1}.
* = carbonaceous chondrites.

In Table 11.1 we summarize the results of the Rb–Sr dating of the various types of meteorites. It is evident that a major chemical differentiation occurred for all the meteoritic types approximately $4·5 \pm 0·2$ b.y. ago (or $4·3 \pm 0·2$ b.y. if $\lambda = 1·47 \times 10^{-11}$ y^{-1}). The Rb–Sr iron meteorite data give no evidence of the great ages reported for these objects by K–Ar dating (Stoenner and Zähringer, 1958).

Murthy and Compston (1965) and Shields *et al.* (1966) studied individual chondrules separated from single meteorites but found results which were scattered on an isochron plot. Shields *et al.* (1966) considered their scatter to result from terrestrial contamination.

As may be seen from Table 11.1, two ages occur at about $3 \cdot 7$–$3 \cdot 8$ b.y. Compston *et al.* (1965) reported this age for the Bishopville aubrite and suggested that an "important chemical fractionation in at least certain meteorites has occurred as late as $3 \cdot 7$ b.y., or has extended for about 1 b.y. from the $4 \cdot 7$ b.y. datum". Burnett and Wasserburg (1967) found a similar result for the Kodaikanal iron meteorite and adopted similar explanations.

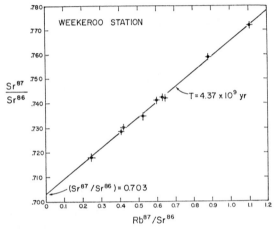

Fig. 11.3. Rb–Sr isochron for silicate nodules from Weekeroo Station iron meteorite. (After Burnett and Wasserburg, 1967.)

Pb^{207}–Pb^{206}

If, at some given time in the past, the various meteorites contained lead of a single isotopic composition, and if at this time uranium and lead in varying amounts relative to each other were segregated into different meteoritic bodies, then present-day analysis of the ratios Pb^{207}/Pb^{204} and Pb^{206}/Pb^{204} would yield a linear array in a plot of Pb^{207}/Pb^{204} against Pb^{206}/Pb^{204}. That is, if the various U–Pb systems have remained closed since the initial event at time t_0. This follows from equation (6.4) of Chapter 6 when $t = 0$. The slope of such a line is

$$\frac{1}{137 \cdot 8} \frac{e^{\lambda_{235} t_0} - 1}{e^{\lambda_{238} t_0} - 1},$$

where $1/137 \cdot 8$ is the ratio U^{235}/U^{238} today. The results of such an analysis by Patterson (1955) gave an age of $4 \cdot 55 \pm 0 \cdot 07$ b.y. for the various meteorite bodies. The iron and troilite (FeS) phases of meteorites usually contain negligible amounts of uranium and consequently their lead frequently has the primeval composition referred to above, and of course falls at one end of the isochron which is the term employed to describe the type of straight line found in plots of Pb^{207}/Pb^{204} versus Pb^{206}/Pb^{204}. Subsequent calculations along these lines by Murthy and Patterson (1962) incorporating more data (Fig. 11.4) yielded essentially the same value as above. Further discussion of meteorite lead is given in Chapter 12.

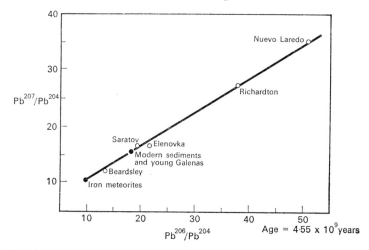

FIG. 11.4. Meteoritic lead isotope isochron. (After Murthy and Patterson, 1962.)

The lead isotope analyses are not easily carried out on meteorites because of their low lead content. In the stones the concentration of lead is generally ~ 1 ppm. In the metallic phase of the irons the concentration is at least as low and is quite uniform ($\sim 0 \cdot 17$ ppm). The troilite phase in the irons is far more variable in lead concentration, ranging from 2 ppm to 60 ppm. It is therefore extremely important to minimize contamination.

Re–Os

Re^{187} decays with the emission of a low-energy β-particle to give Os^{187}. Laboratory measurement of the half-life is difficult and attempts have been made to determine it from measurements of the Os^{187}/Os^{186} ratios and the

Re^{187} contents of minerals whose ages may be inferred from other techniques. A value of $t_{\frac{1}{2}} = 4\cdot3 \pm 0\cdot5 \times 10^{10}$ y has been recently obtained by Hirt *et al.* (1963) in this manner.

In principle it should be possible to use an analysis by this technique which would be analogous with the whole rock Rb–Sr method. In this instance, systems originating simultaneously with the same initial Os^{187}/Os^{186} ratio should, when now analysed, yield points on a graph of Re^{187}/Os^{186} versus Os^{187}/Os^{186} which delineate a straight line. The experimental results for iron meteorites have been summarized by Anders (1963). The best straight line through the points gives an age of $4\cdot0 \pm 0\cdot8$ b.y. when the above-mentioned half-life is used. It is clear that only an imprecise value can be deduced from present data. A major difficulty seems to be that rhenium and osmium behave similarly geochemically in iron meteorites so that little variation in the Re^{187}/Os^{186} ratio is found and consequently the narrow cluster of points on the isochron plot gives a poor basis for the calculation of a straight line.

11.3. Summary

From the foregoing review of meteorite data it seems possible to make a reasonable estimate of the age of the meteorites. The lead and rubidium-strontium techniques date a time of geochemical separation of meteorite phases. They date a separation of lead from uranium or strontium from rubidium. Results of lead analysis strongly point to such a differentiation event occurring about $4\cdot6$ b.y. ago. Essentially the same conclusion may be drawn from the Rb–Sr figures, although the annoying 6% *minimum* uncertainty in the rubidium half-life prevents clear comparison. The Rb–Sr data also suggest a later event at about $3\cdot8$ b.y. Concerning the Re–Os measurements one can merely say that they show broad agreement with the bulk of the Rb–Sr and Pb–Pb data.

The ages involving the gaseous isotopes Ar^{40} and He^4 indicate the time when the meteorites were sufficiently cool to retain these daughters. The large cluster of ages between $4\cdot0$ and $4\cdot8$ b.y. resulting from both methods strongly indicates a severe heating episode at some time in that interval and it would seem reasonable to identify such an episode with that in which the uranium, lead, rubidium and strontium elements were fractionated. The younger K–Ar and U–He ages must be taken to indicate the results of at least one later disturbance—a passage in flight close to the sun is one suggestion. Until more has been discovered about the meteorites, however, we will remain uncertain. The analysis of different mineral phases has only recently begun but it should shed a great deal more light on the problems.

11.4. The Age of the Elements

If the earth were "formed" about $4 \cdot 6$ b.y. ago it is interesting to question just how old were the atoms from which the earth itself was made? Was the earth consolidated a few hours, a few years or a few billion years after the elements comprising it were formed? No definite answers are known to these questions but much work has been done on them.

One attack, first adopted by Rutherford, is to consider the element uranium. The decay constant of U^{238} is $1 \cdot 54 \times 10^{-10} \, y^{-1}$ and that of U^{235} is $9 \cdot 72 \times 10^{-10} \, y^{-1}$. The present ratio of U^{238}/U^{235} is $137 \cdot 8$ as measured for uranium extracted from deposits or minerals anywhere in the world. Then we may write

or

$$\left(\frac{U^{238}}{U^{235}}\right)_{now} = \left(\frac{U^{238}}{U^{235}}\right)_{init} \frac{e^{-\lambda t}}{e^{-\lambda' t}} = \left(\frac{U^{238}}{U^{235}}\right)_{init} e^{-(\lambda - \lambda')t}$$

$$t = \frac{1}{\lambda' - \lambda} \ln \frac{(U^{238}/U^{235})_{now}}{(U^{238}/U^{235})_{init}}, \qquad (11.1)$$

where the decay constants of U^{238} and U^{235} are λ and λ' respectively. Setting

$$\left(\frac{U^{238}}{U^{235}}\right)_{now} = 137 \cdot 8,$$

its measured value, and assuming

$$\left(\frac{U^{238}}{U^{235}}\right)_{init} = 1,$$

then we have $t \simeq 6$ b.y. That is, under the above assumption, it is estimated that the elements were formed approximately 6 b.y. ago. Clearly this can only be regarded as a tentative calculation. If we assumed the initial ratio of U^{238}/U^{235} were 2, for instance, then we would find that the elements formed approximately $5 \cdot 2$ b.y. ago. Thus an uncertainty of a factor of 2 in our knowledge of the initial relative abundances of the uranium isotopes produces an uncertainty of 800 m.y. in our estimate of the age of the elements.

A more accurate estimate would be obtained from the study of an isotopic ratio which varied much more strongly with time. In such an instance an uncertainty of a factor of 2 in the isotopic ratio would mean but a small uncertainty in time. One might therefore look for a radioactivity of much shorter half-life than that of either uranium isotope. The immediate objection to this is that such an activity, formed over 4·6 b.y. ago, would long since have become extinct. With such a system, therefore, it is necessary to study the *daughter* isotope of the decay and its variations in abundance.

11.5. I–Xe **Dating**

From nuclear physical experiments it is known that the isotope of iodine, I^{129}, can exist, with a half-life of 16·4 m.y. Having such a brief half-life it is of course not found in nature—naturally occurring iodine having the one isotope I^{127}, which is stable. An isotope of the rare gas xenon, Xe^{129}, is formed by the β^- decay of I^{129}. Thus if the meteorites were formed within a few multiples of 16·4 m.y., of the time of cessation of I^{129} synthesis, they would trap I^{129} along with the stable I^{127}. This trapped I^{129} would quickly decay to Xe^{129}. The first observation of excess Xe^{129} presumed to be from this source was made by Reynolds (1960) in an analysis of the Richardton chondrite. I–Xe ages based on this effect, a method proposed by Brown (1947), have usually been calculated on one of two models— the "first five minutes" hypothesis of nuclear creation and the model of Wasserburg *et al.* (1960).

(a) FIRST FIVE MINUTES

This is based on a theory, strongly supported by Gamow (Alpher *et al.*, 1948), of the creation of the elements in some explosively short space of time. The universe is envisaged as at its beginning being a highly concentrated mass of radiation and elementary matter. Within this mixture the various elements were synthesized before the agglomeration burst explosively apart. At some stage after the explosion the solar system and other features of the universe evolved. The I–Xe method may be used in an endeavour to estimate the time elapsing between the end of the nuclear synthesis and the consolidation of the earth, the latter being assumed contemporaneous with the formation of the meteorites. In Fig. 11.5 we give a picture of the supposed history of the universe on this model. During the "first five minutes" I^{129} and I^{127} are generated, as are the various isotopes of xenon. Once the initial creative episode ceases the I^{129} decays without replenishment into Xe^{129}. This decay is assumed to go on for a length of time t until formation of the solar system including the earth and meteoritic matter takes place. A meteorite body is assumed to trap some

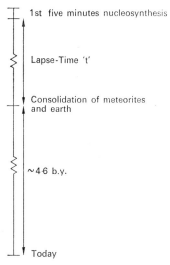

1st five minutes nucleosynthesis

Lapse-Time 't'

Consolidation of meteorites and earth

~4·6 b.y.

Today

FIG. 11.5. "First five minutes" model for I–Xe dating.

of the primeval xenon and some iodine. If t is not too many units of 16·4 m.y., then the trapped iodine will consist of the stable I^{127} and some remaining I^{129}. The latter subsequently decays to Xe^{129} until, when we analyse a meteorite today, we find in it no I^{129}, only I^{127}. However, the xenon in the meteorite will be composed of xenon of primeval composition plus the Xe^{129} into which the trapped I^{129} has decayed. The theoretical basis of the calculation is as follows:

Suppose that at the end of the nucleosynthesizing event the ratio of I^{129}/I^{127} is $(I^{129}/I^{127})_0$. Then after a time t has elapsed the ratio will have fallen to

$$\left(\frac{I^{129}}{I^{127}}\right)_{\text{inc}} = \left(\frac{I^{129}}{I^{127}}\right)_0 e^{-\lambda t}$$

where λ is the decay constant of I^{129}.

$(I^{129}/I^{127})_{\text{inc}}$ will be the isotopic composition of the iodine incorporated into the solid objects of the solar system. Since all the trapped I^{129} decays to Xe^{129} we can then write

$$\left(\frac{I^{129}}{I^{127}}\right)_{\text{inc}} = \frac{Xe^{129}}{I^{127}}$$

where Xe^{129} is the excess xenon-129 in the meteorite over and above the amount of this isotope trapped at meteorite formation. The age measurement thus consists in determining both the I^{127} concentration in the meteorite and the excess Xe^{129}. The latter is often found by comparing the isotopic composition of the xenon extracted from the meteorite with that of atmospheric xenon, the anomaly at mass 129 being assumed due to decay of I^{129}. The calculation is completed by writing

$$\frac{Xe^{129}}{I^{127}} = \left(\frac{I^{129}}{I^{127}}\right)_{\text{inc}} = \left(\frac{I^{129}}{I^{127}}\right)_{0} e^{-\lambda t},$$

which gives

$$t = \frac{1}{\lambda} \ln \left\{ \left(\frac{I^{129}}{I^{127}}\right)_{0} \cdot \frac{I^{127}}{Xe^{129}} \right\}. \tag{11.2}$$

Hence the time t, which is usually called a "lapse time", may be calculated if some value for $(I^{129}/I^{127})_{0}$ is assumed.

(b) WASSERBURG, FOWLER, HOYLE MODEL

This is the more commonly used model, on which it is assumed that the universe is considerably older than 4·6 b.y. and that element formation is taking place over long periods of time within large stars. Newly formed material is every so often fired out into interstellar space during catastrophic events in the stars. The theory is simply illustrated in Fig. 11.6. It can easily be shown that at the close of nucleosynthesis the ratio of the iodine isotopes is given by

$$\left(\frac{I^{129}}{I^{127}}\right)_{0} = \frac{K_{129}\tau}{K_{127}T},$$

where K_{127}, K_{129} are the respective production rates of I^{127}, I^{129} within the star; $\tau = 1/\lambda =$ the mean life of I^{129}; $T =$ the period of nucleosynthesis. This iodine along with the other elements exists for a time t outside its formative star until it is incorporated in the meteoritic system. During this period t the iodine isotopic ratio will have fallen to

$$\left(\frac{I^{129}}{I^{127}}\right)_{inc} = \left(\frac{I^{129}}{I^{127}}\right)_0 e^{-\lambda t}.$$

This is the composition of the iodine trapped in the meteorite matter. The calculation proceeds then exactly as in the previous theory, so that

$$t = \frac{1}{\lambda} \ln \left\{ \left(\frac{I^{129}}{I^{127}}\right)_0 \frac{I^{127}}{Xe^{129}} \right\} = \frac{1}{\lambda} \ln \left\{ \frac{K_{129}}{K_{127}} \frac{\tau}{T} \frac{I^{127}}{Xe^{129}} \right\}. \qquad (11.3)$$

A value of $(I^{129}/I^{127})_0 = 1 \cdot 25 \times 10^{-3}$ is widely adopted on this model, an error of a factor of 10 in this ratio producing an error of ± 55 m.y. in the calculated lapse time. Table 11.2 shows some of the values of t which

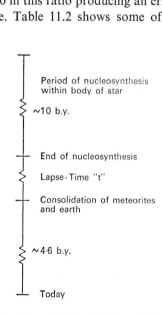

Period of nucleosynthesis within body of star

~10 b.y.

End of nucleosynthesis

Lapse-Time "t"

Consolidation of meteorites and earth

~4·6 b.y.

Today

FIG. 11.6. Wasserburg, Fowler, Hoyle model for I–Xe dating.

TABLE 11.2. I–Xe LAPSE TIMES

(Modified from Anders, 1964)

Meteorite	W.F.H. lapse time (m.y.)	Correlated I lapse time (m.y.)
Abee	47	51·5 ± 2
Indarch	77	
St Marks	52	
Beardsley	254	
Richardton	97	51·5 ± 2
Bruderheim	107	34·3 ± 6
Renazzo		66 ± 6
Murray	> 128	
Sardis Troilite	238	

W.F.H. = Wasserburg, Fowler and Hoyle.
For references to W.F.H. column see Anders (1963). The correlated
lapse times are from Reynolds (1963) and Reynolds and Turner (1964).

have been determined on this theory. The results indicate that the meteorites, and by inference the earth and probably the whole solar system, were formed of the order of 100 m.y. after the synthesis of their constituent elements. As should be obvious, the calculations involve a number of assumptions. One problem is to decide on exactly what is primeval xenon so that the magnitude of the anomaly at mass 129 can be accurately known. The choice of atmospheric xenon as primeval is obviously a makeshift one, especially in the light of the observation by Reynolds (1960) that meteoritic xenon is anomalous at every mass number with respect to atmospheric xenon. Another problem is posed by the question of the correlation of the excess Xe^{129} with the iodine in the meteorite. While Jeffery and Reynolds showed that excess Xe^{129} was in an iodine-bearing phase, it was then found (Goles and Anders, 1962; Reynolds, 1963) that there are several different iodine sites, with the Xe^{129} being correlated with iodine in sites which are more retentive of xenon. Fish and Goles (1962) therefore developed a method designed to overcome both the uncertainty concerning primeval xenon composition and the iodine–xenon correlation problem which results in the modified ages given in Table 11.2. In this method neutron irradiation of the meteorite is used to convert some of the I^{127} into Xe^{128} by the nuclear reaction $I^{127}(n,\gamma\beta)Xe^{128}$ (Jeffery and Reynolds,

1961). Thus the xenon in the meteorite consists of a primeval component together with a so-called "radiogenic" component comprised of the excess Xe^{129} formed by the decay of the now-extinct I^{129} and the Xe^{128} formed in the neutron bombardment of the iodine in the meteorite. If the excess Xe^{129} is correlated entirely with the iodine sites, as is the neutron-produced Xe^{128}, then it is straightforward to show that there will be a linear relation between the ratios Xe^{129}/Xe^{132} and Xe^{128}/Xe^{132} measured in the gas fractions released from a meteorite heated in a step-wise fashion. The slope of a straight line obtained in such an experiment gives a measure of the isotopic ratio $(I^{129}/I^{127})_{inc}$ of the iodine trapped in the meteorite at the commencement of xenon retention. The reciprocal of this ratio may then be substituted in equation (11.3) as the value of I^{127}/Xe^{129} and the lapse time may be calculated if $(I^{129}/I^{127})_0$ is assumed.

A most striking application of this method was made by Hohenberg, Podosek and Reynolds (1967), who showed that the Xe^{129}/Xe^{132}, Xe^{128}/Xe^{132} data from stepwise heating experiments with ten different chondrites all fell on a single straight line for gas fractions extracted by heating above 1100°C. This demonstrated that the $(I^{129}/I^{127})_{inc}$ ratio was essentially identical for all the chondrites examined, which implies very strongly that the different meteorites (or at least the chondrules in them) all cooled to xenon-retention temperatures at about the same time, with an uncertainty in this simultaneity of only 2 m.y. The single lapse time calculated for all these bodies would be about 60 m.y. on the Wasserburg, Fowler, Hoyle model.

CHAPTER 12

THE AGE OF THE EARTH

12.1. Historical

At the end of the nineteenth century the question of the age of the earth was a source of many lively debates. On the one hand, there were numerous estimates of 100 m.y. or less, while many scientists favoured values much greater than this. The speculations leading to the smaller ages were mainly drawn from the province of physics. From a consideration of the duration of the sun's heat Helmholtz calculated a possible age of 22 m.y. Kelvin, supposing the earth to have been initially molten, estimated that it would require 20–400 m.y. to cool to the present low surface heat flow. A figure of 57 m.y. was found by G. H. Darwin in his investigations of the possibility of the separation of the moon from the earth. The rate of accumulation of sodium in the oceans was used by the geologist Joly to calculate an age of 80–90 m.y. for the oceans.

Many geologists, however, were not impressed by these arguments and considered the age values to be gross underestimates. Charles Darwin in his Origin of Species said:

> Sir W. Thompson (Lord Kelvin) concludes that the consolidation of the crust can hardly have occurred less than 20 or more than 400 m.y. ago, but probably not less than 98 or more than 200 m.y. These very wide limits show how doubtful the data are; and other elements may have hereafter to be introduced into the problem. Mr Croll estimates that about 60 m.y. have elapsed since the Cambrian period, but this, judging from the small amount of organic change since the commencement of the Glacial epoch, appears a very short time for the many and great mutations of life, which have certainly occurred since the Cambrian formation, and the previous 140 m.y. can hardly be considered as sufficient for the development of the varied forms of life which already existed during the Cambrian period.

In 1893 Reade, from a study of rates of sedimentation, concluded that 600 m.y. had elapsed since the beginning of the Cambrian, while Goodchild found 700 m.y. from a similar calculation.

Kelvin's solution of the thermal problem turns out on close scrutiny to involve merely the cooling of a very thin outer layer of the earth. The very great preponderance of the earth's initial heat remained stored inside below a depth of about 250 km. It was pointed out by Perry and Heaviside that if this massive heat supply were tapped in some way, then considerably greater ages could be obtained for the earth. Yet there remained an even greater flaw. There was, as Darwin surmised, another element which had to be introduced. In succession, Becquerel discovered natural radioactivity in 1896; Pierre Curie and Laborde observed in 1903 that radium maintains a higher temperature than its surroundings; three years later R. J. Strutt (later Lord Rayleigh) discovered that radioactivity characterizes all rocks and addressed the Royal Society "On the distribution of radium in the Earth's crust, and on the Earth's internal heat". The doom of all the old estimates was sealed.

In 1904 Rutherford suggested that the ages of rocks might be calculated from the rate of accumulation of helium in radioactive minerals. The suggestion was taken up by Boltwood and Rayleigh, and Boltwood also initiated the U–Pb method after proposing that lead was the final product of uranium decay. However, subsequent progress was relatively slow, principally because of difficulties with the comparative ease with which helium escapes from many minerals and equally because time had to be allowed for Aston and Dempster to develop the mass spectrometer. By 1931 Holmes was able to conclude that the age of the earth "exceeds 1460 m.y., is probably not less than 1600 m.y. and is probably much less than 3000 m.y.". The lower limit was found from a consideration that the highest reliable lead age was 1460 m.y. for the Keystone uraninite from Black Hills, South Dakota. For the upper limit, Holmes revised a calculation first made by H. N. Russell (1921) in which it was assumed that all the lead in igneous rocks was of radioactive origin. The mean concentration by weight of lead in igneous rocks was taken as $7 \cdot 5$ ppm and the concentration by weight of uranium was taken by Holmes to be $11 \cdot 4$ ppm including the uranium equivalent of thorium. So we have for the atomic daughter to parent ratio $Pb/U = (7 \cdot 5/11 \cdot 4) \times (238/207) = 0 \cdot 76$. Hence, substituting this ratio in our fundamental age formula, equation (2.2), we get $t = 3700$ m.y. for the age of the earth. Since, of course, not all the lead is of radioactive origin this of necessity sets an upper limit. Furthermore, at this time it was thought that Pb^{207} was not radiogenic, so Holmes corrected his lead concentration downward and finally took approximately 3000 m.y. as a reasonable upper limit (Holmes, 1931).

COLLEGE OF THE SEQUOIAS
LIBRARY

12.2. Modern Estimates

In principle the modern means of estimating the earth's age differ but slightly from those reviewed by Holmes in 1931. Thus we still find the minimum age of the earth from the age of the oldest terrestrial rocks. And we obtain still greater values from considerations of the evolution of lead isotopes in the earth. A new development has been the introduction into the discussions of data from the meteorites.

If we examine Dearnley's histogram in Fig. 9.1 we can immediately see the tremendous change which has taken place in the geochronological picture since 1931. Whereas the oldest reliable lead age then was $1 \cdot 46$ b.y., we now see that this value would be roughly near the middle of the present mineral age distribution. The greater part of the surface of the earth's crust evidently lies in the age range 0–2·8 b.y. However, while the age distribution is sharply reduced at 2·8 b.y., it does not stop there abruptly and there are undoubtedly rocks now present at the earth's surface of considerably greater age than this. Table 12.1 summarizes evidence of the existence of a continental crust on the earth over 3 b.y. ago.

TABLE 12.1. MINERAL AND WHOLE ROCK AGES GREATER THAN 3·0 b.y.

Location	Refs.	Method	Mineral	Age (b.y.)
Kola Peninsula, U.S.S.R.	1	K–Ar	Biotite	3·46
Ukraine, U.S.S.R.	2	K–Ar	Biotite	3·05
Swaziland	3	Rb–Sr*	Whole rock	3·07 ± 0·06
				3·44 ± 0·30
Transvaal, S. Africa	4	Rb–Sr*	Whole rock	3·20 ± 0·07
Congo	5	Rb–Sr*	Microcline	3·52 ± 0·18
Minnesota, U.S.A.	6	U–Pb	Zircon	≳ 3·3
Montana, U.S.A.	7	U–Pb	Zircon	≳ 3·1

$*\lambda = 1 \cdot 39 \times 10^{-11}$ y^{-1}.

1. Polkanov and Gerling (1961); 2. Vinogradov and Tugarinov (1961); 3. Allsopp *et al.* (1962); 4. Allsopp (1961); 5. Ledent *et al.* (1962); 6. Catanzaro (1963); 7. Catanzaro and Kulp (1964).

Russian investigations have shown that both the Baltic and Ukrainian shields contain rocks of extreme age. Outcropping over a small area in the north-east part of the Kola Peninsula are granites, pegmatites, migmatites and gneisses giving K–Ar ages in the range 3·0–3·5 b.y. Polkanov and Gerling (1961) suggested that there was evidence of two orogenic cycles during this time: the Upper Katarchean $3 \cdot 0$–$3 \cdot 1$ b.y. ago and the Lower Katarchean $3 \cdot 25$–$3 \cdot 5$ b.y. ago. Biotites from migmatites in the area of the Voronezh River gave $3 \cdot 44$ and $3 \cdot 48$ b.y. by K–Ar dating. Moving from the Baltic shield, Vinogradov and Tugarinov (1961) reported a K–Ar age of $3 \cdot 05$ b.y. on biotite from xenoliths of meta-amphibolite in granites in the middle Dnieper region of the Ukraine.

Similarly, great ages have been found for the African continent. Allsopp *et al.* (1962) analysed five whole rocks from the "Old Granites" of central Transvaal, South Africa, by the Rb–Sr method and obtained an extremely good isochron whose slope indicated that these Old Granites were intruded $3 \cdot 2 \pm 0 \cdot 07$ b.y. ago. The isochron is shown in Fig. 5.2. These granites were emplaced into rocks now comprising the basement, prominent among the latter being metavolcanics, serpentines, banded iron-stones and schists. Evidently these would be older than $3 \cdot 2$ b.y. Confirmatory evidence for these ages is provided by the results obtained by Nicolaysen *et al.* (1962) from the Dominion Reef conglomerates which lie on the eroded surface of the "Old Granites". U–Pb whole rock analyses clearly indicated that the materials comprising the conglomerate were $3 \cdot 1 \pm 0 \cdot 1$ b.y. old.

To the east of these rocks, Allsopp *et al.* (1962) found another locality of ancient granites in Swaziland. The average of four whole-rock samples of so-called G4 granite was $3 \cdot 07 \pm 0 \cdot 06$ b.y. if it was assumed that the initial strontium isotope ratio was $(Sr^{87}/Sr^{86})_i = 0 \cdot 710$. Two analyses of the G1 granite, which is supposedly older on geological grounds, yielded $3 \cdot 30 \pm 0 \cdot 32$ b.y. and $3 \cdot 58 \pm 0 \cdot 51$ b.y. for an average of $3 \cdot 44 \pm 0 \cdot 30$ b.y. However, G1 showed such small Sr^{87} enrichment that the age found was critically dependent on the assumed $(Sr^{87}/Sr^{86})_i$ and it may be said to be dated only within very wide limits. According to Hunter (1957), the G4 granite is intrusive into the Mozaan series which overlies the Insuzi series, so both these series must be older than $3 \cdot 07 \pm 0 \cdot 06$ b.y. Similarly, the Swaziland system, being pre-G1, was tentatively considered to be greater than $3 \cdot 44 \pm 0 \cdot 30$ b.y. in age by Allsopp *et al.* A further indication of the antiquity of this continent was found by Ledent *et al.* (1962) in the basement rocks of the southern Congo. Microcline from a pegmatite was calculated to have a Rb–Sr age of $3 \cdot 52 \pm 0 \cdot 18$ b.y. While we have seen in Chapter 5 that a single Rb–Sr mineral age is necessarily suspect, this occurrence is obviously well worth further investigation.

The North American continent has been shown to be considerably older than 3 b.y. by the U–Pb dating of zircons. Catanzaro (1963) published the

analyses of two zircons from the Morton gneiss and one from the Montevideo gneiss. His data are seen in a concordia plot in Fig. 12.1. The samples
do not fall on the concordia curve and so have discordant U–Pb ages.
Discordancy was not unexpected, however, since it had been earlier shown
by Goldich *et al.* (1961) and Goldich and Hedge (1962) that this area of
south-west Minnesota underwent severe regional metamorphism about
$2 \cdot 5 \pm 0 \cdot 1$ b.y. ago and experienced further but milder igneous activity

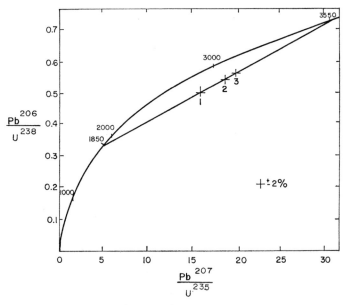

Fig. 12.1. Concordia plot of zircons from Morton and Montevideo Gneisses.
(After Catanzaro, 1963.)

$1 \cdot 8 \pm 0 \cdot 1$ b.y. ago. Referring again to Fig. 12.1 we see that in fact the
discordant zircons lie on a straight line which intersects the concordia
curve at $3 \cdot 55$ b.y. and $1 \cdot 85$ b.y. Interpreting this on the episodic loss
model of Wetherill (1956), as described in Chapter 5, it would be concluded
that these zircons crystallized originally $3 \cdot 55$ b.y. ago and underwent an
episode involving the loss of lead $1 \cdot 85$ b.y. ago. This latter event would
obviously correlate well with the second heating of the area observed by
Goldich and co-workers, and this interpretation is probably correct. The
fact that the severe metamorphism $2 \cdot 5 \pm 0 \cdot 1$ b.y. ago apparently had no

effect on the lead content of the zircons, while surprising, is not entirely without precedent. As Catanzaro noted, it seems easier to remove lead from zircons as they get older, presumably due to the increased amount of radiation damage with time, and the extra $0 \cdot 7$ b.y. of history may have rendered these zircons more susceptible to alteration in the second event. If one endeavours to fit the data to Tilton's diffusion loss model, Catanzaro notes that one obtains a minimum crystallization age of about $3 \cdot 3$ b.y. Whichever interpretation is adopted it seems evident that the zircons are very old. Goldich *et al.* (1961) considered that the Montevideo granite gneiss intruded the "basic complex" which was comprised of two units: (a) gabbroic and dioritic gneiss and (b) garnetiferous quartz–diorite gneiss. If the zircons of the Morton and Montevideo gneisses were not derived from this basic complex, the latter may eventually give still older ages.

Further west, in the Beartooth Mountains of Montana, Catanzaro and Kulp (1964) found more evidence of the great age of the North American continent. Seven detrital zircons from Precambrian metasedimentary rocks gave grossly discordant U–Pb ages, the Pb^{207}/Pb^{206} ages ranging from $2 \cdot 58$ to $3 \cdot 09$ b.y. While the picture was far from clear, these results strongly suggested an age greater than 3 b.y. and Catanzaro and Kulp, after an exhaustive discussion, proposed a minimum age for the zircons of $3 \cdot 1$ b.y. These authors also referred in their paper to K–Ar ages obtained in their laboratory on biotites from the area which were of the order $3 \cdot 1$ b.y.

Each continent, except Antarctica, is thus successively giving evidence of being much older than 3 b.y. In the South African and North American situations, there is evidence of older formations than those dated at more than 3 b.y. If we consider that the Lower Katarchean orogenic cycle was likely not the earth's first such episode, then it seems that a continental crust was in existence $3 \cdot 7$ b.y. ago and perhaps 4 b.y. ago. Some parts of the continental crust of the earth are thus being shown to be measurably older as the number of geochronometric data swells, and the gradual appreciation of this fact must have an increasing influence on general theories of the earth's formation and evolution.

12.3. Estimates from Terrestrial Leads

Following Aston's pioneering application of mass spectrometry to lead contained in radioactive minerals, Nier (1938, 1939 and Nier, Thompson and Murphey, 1941) examined the isotopic composition of common lead in various non-radioactive minerals. These researches form a land-mark in geochronology. Nier observed that the proportions of Pb^{208}, Pb^{207} and Pb^{206} relative to Pb^{204} decreased as the age of the samples increased. He proposed that these variations resulted quite simply from the various leads being removed at differing times from a source area which contained uranium and thorium. Clearly, the longer a lead sample was in this radio-active environment, the more radiogenic would be the composition of its lead. The mathematical expression of this process was given by Gerling (1942), Holmes (1946, 1947 and 1949) and Houtermans (1946 and 1947), and Nier's data were used by these authors to calculate the age of the earth. Later analyses of a similar type were made by Bullard and Stanley (1949), Alpher and Herman (1951) and Collins et al. (1953), who added new analytical data. The computations are essentially based on the four following assumptions:

(a) the isotopic composition of lead was uniform throughout the earth at an early stage in its formation;
(b) at this stage U/Pb and Th/Pb ratios were frozen to characteristic values in various regions and varied subsequently only by reason of radioactive decay;
(c) each lead sample was derived from one such area and has sub-sequently been unaltered in isotopic constitution;
(d) the time at which the lead was removed from its parental environment to form a lead ore may be determined either from stratigraphic argu-ments or from radiometric analyses on associated minerals.

If these requirements are satisfied, then the isotope ratios of lead found in minerals such as galena will be described by equation (6.4) of Chapter 6. That is,

$$\frac{y - b_0}{x - a_0} = \frac{1}{137 \cdot 8} \frac{(e^{\lambda_{235} t_0} - e^{\lambda_{235} t})}{(e^{\lambda_{238} t_0} - e^{\lambda_{238} t})} \tag{12.1}$$

where $137 \cdot 8 = U^{238}/U^{235}$, and x and y are the Pb^{206}/Pb^{204} and Pb^{207}/Pb^{204} ratios which would be measured today for the galena formed at time t. Hence, if we have a set of galenas whose ages are presumed known and whose lead isotope ratios have been measured we can apply the method of least squares to this equation to find the values of the parameters a_0, b_0 and t_0 which best fit these data. The results obtained by various workers are shown in Table 12.2, and give an average estimated age for the earth

TABLE 12.2. SOME ESTIMATES OF THE AGE OF THE EARTH UP TO 1953

Author	t_0 (b.y.)	a_0	b_0
Holmes (1947)	3·35	10·95	13·51
Houtermans (1947)	2·9	11·52	14·03
Bullard and Stanley (1949)	3·29	11·86	13·86
Collins, Russell and Farquhar (1953)	3·5	11·83	13·55

of 3·26 b.y. and the mean values $a_0 = 11·54$, $b_0 = 13·74$ for the initial lead isotope ratios in the earth when the U/Pb and Th/Pb ratios were frozen.

12.4. Estimates Combining Meteoritic and Terrestrial Leads

The last of the above estimates was published in 1953, the year in which the first isotopic analyses of lead from an iron meteorite were announced by Patterson (1953). Thenceforward the lead isotope ratios in meteorites have shared the stage equally with terrestrial lead ratios in determinations of the earth's age. Patterson, Tilton and Inghram (1955) proposed to solve equation (12.1) for t_0 very simply by setting a_0 and b_0 equal to the Pb^{206}/Pb^{204} and Pb^{207}/Pb^{204} ratios measured in the troilite (FeS) phase of iron meteorites and taking x and y to be the Pb^{206}/Pb^{204} and Pb^{207}/Pb^{204} isotopic composition of "modern lead" for which $t = 0$. Equation (12.1) then becomes

$$\frac{y - b_0}{x - a_0} = \frac{1}{137·8} \frac{(e^{\lambda_{235} t_0} - 1)}{(e^{\lambda_{238} t_0} - 1)}. \qquad (12.2)$$

Since x, y, a_0 and b_0 are then known, we can solve this equation for t_0. This calculation still involves assumptions (a) to (d) defined earlier. The additional assumption made is that the lead isotope ratios now found in the troilite phase of iron meteorites are those values characterizing the earth's lead at the time of the earth's formation. This is obviously a critical assumption and unfortunately it has not yet been clearly validated. However, as we have seen in Chapter 11 the U/Pb ratio in troilite is *usually* so low that its lead isotope ratios do not change measurably with time. Furthermore, such troilite lead is the least radiogenic of all lead, so the assumption is at least reasonable. The averages of the isotope ratios of lead from the Canyon Diablo and Henbury troilites were found to be

$a_0 = 9 \cdot 46$ and $b_0 = 10 \cdot 29$. For samples of "modern lead", that is lead removed in geologically recent times from its parental uranium- and thorium-rich environment, Patterson *et al.* (1955) used the lead in recent volcanic rocks and lead ores. This lead they considered to have an isotopic constitution falling within the range $x = 18 \cdot 07$–$18 \cdot 95$ and $y = 15 \cdot 40$–$15 \cdot 76$. These values, when combined with the meteoritic a_0 and b_0 then gave $t_0 = 4 \cdot 5$ b.y. approximately. Thus t_0, when estimated in this combined meteoritic–terrestrial approach, gave an age for the earth which was greater than the immediately preceding purely terrestrial estimates by

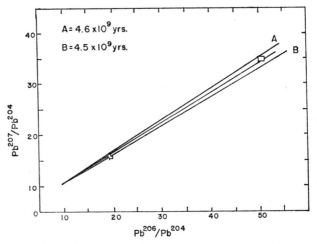

Fig. 12.2. Patterson's original meteorite isochron. (After Patterson, 1956.)

approximately one billion years, i.e. by about 30%. These investigators, however, concluded their paper with the caution: "It should be viewed with considerable skepticism until the basic assumptions that are involved in the method of calculation are verified."

In the following year, Patterson (1956) attempted a more precise estimate and to the two iron meteorite analyses already used he added measurements of the lead isotope ratios of three stone meteorites, Forest City, Modoc and Nuevo Laredo. These data are shown in Fig. 12.2 on a graph of Pb^{207}/Pb^{204} vs. Pb^{206}/Pb^{204}, where they may be seen to define a straight line quite well. But this is what we would expect if the meteorites had a common initial lead (a_0, b_0) and had different U/Pb ratios among them-

selves which were only perturbed by radioactive decay until very recent
times. For such leads are described by equation (12.2), which shows that a
graph of y vs. x would give a straight line passing through the primeval
lead point (a_0, b_0) and of slope

$$R = \frac{1}{137 \cdot 8} \frac{(e^{\lambda_{235} t_0} - 1)}{(e^{\lambda_{238} t_0} - 1)}.$$

Solving this equation for t_0, Patterson obtained the age $4 \cdot 55 \pm 0 \cdot 07$ b.y.
Since our earth's so-called "modern lead" lay on this line within experimental error, Patterson concluded that the age of the earth was also
$4 \cdot 55 \pm 0 \cdot 07$ b.y. These elegant meteoritic studies were immediately
recognized for their great worth and this value for the age of the earth has
achieved almost universal currency.

Recently two more studies have been published, both of which use the
troilite lead isotope data to fix a_0 and b_0 but eschew the stone meteorite
figures. Ostic et al. (1963) combined the troilite primeval lead values with
the isotope ratios of nine "conformable" galenas. We have seen earlier
in Chapter 6 that this type of galena, as defined by Stanton, is supposed to
contain lead which has had a particularly simple history. Ostic et al.
proposed that the lead in each of the nine galenas developed in a uniform
source area within the earth, was removed at some time to form galena
and was subsequently undisturbed. The development of the lead isotope
ratios in the uniform source would be described by equations (6.1), (6.2)
and (6.3) of Chapter 6, and so the age of a galena formed from such material at time t would be given by both

$$t_{206} = \frac{1}{\lambda_{238}} \ln \left\{ e^{\lambda_{238} t_0} - \frac{x - a_0}{a V} \right\} \tag{12.3}$$

and

$$t_{207} = \frac{1}{\lambda_{235}} \ln \left\{ e^{\lambda_{235} t_0} - \frac{y - b_0}{V} \right\}, \tag{12.4}$$

where $V = U^{235}/Pb^{204}$ and $a = 137 \cdot 8$.

Ostic et al. accordingly carried out a least squares analysis of the galena
data (i.e. x and y) which chose t_0 and V to give the most concordant pairs
of t_{206} and t_{207}. To do this they minimized the sum $\Sigma W(t_{206} - t_{207})^2$,
where the W's were the appropriately chosen weighting factors. The greatest
virtue of this approach is that the ages of the galenas do not therefore need

to be known. With this calculation, assuming $a_0 = 9 \cdot 56$ and $b_0 = 10 \cdot 42$, the magnitudes of t_0 and V which were found to minimize the sum just described were $t_0 = 4 \cdot 53$ b.y. (s.d. $= 0 \cdot 03$ b.y.) and $V = 0 \cdot 067$. In addition to assumption (a) defined earlier in the chapter, this method involves the supposition that all these conformable galenas drew their lead from areas which all had the same U/Pb ratio. Using another variant these same investigators found $t_0 = 4 \cdot 54 \pm 0 \cdot 02$ b.y. These values for the age of the earth thus agree closely with the earlier results of Patterson and his co-workers.

The second recent estimate is due to Tilton and Steiger (1965), who concluded that the earth may be approximately 200 m.y. older than earlier thought. Their approach in principle differs but little from the earlier one of Patterson *et al.* (1955) which we have described. They similarly proposed to solve equation (12.1) for t_0, using troilite values for a_0 and b_0. However, instead of using "modern lead" for x and y, and setting $t = 0$, they used the composition of lead from galena from Manitouwadge, Ontario, for x and y and therefore made t in equation (12.1) equal to the age of this galena. This was achieved by U–Pb and Rb–Sr dating of nearby granites which indicated a time of emplacement of 2700 ± 100 m.y. With the quantities a_0, b_0 and t put equal to $9 \cdot 54$, $10 \cdot 27$ and 2700 m.y., Tilton and Steiger solved equation (12.1) and observed that $t_0 = 4 \cdot 75 \pm 0 \cdot 05$ b.y.

12.5. Commentary

In Table 12.3 we have summarized the various combined meteoritic–terrestrial estimates of the earth's age based on theories of the evolution of lead isotope ratios. We note that, while the spread is small, so is that in Table 12.2, yet radically new information from the meteorites caused a

TABLE 12.3. ESTIMATES OF THE AGE OF THE EARTH COMBINING METEORITIC AND TERRESTRIAL LEAD DATA

Author	t_0 (b.y.)	a_0	b_0
Patterson, Tilton and Inghram (1955)	$4 \cdot 5$	$9 \cdot 46$	$10 \cdot 29$
Patterson (1956)	$4 \cdot 55 \pm 0 \cdot 07$	$9 \cdot 50$	$10 \cdot 36$
Ostic, Russell and Reynolds (1963)	$4 \cdot 53 \pm 0 \cdot 03$	$9 \cdot 56$	$10 \cdot 42$
Tilton and Steiger (1965)	$4 \cdot 75 \pm 0 \cdot 05$	$9 \cdot 54$	$10 \cdot 27$
Ulrych (1967)	$4 \cdot 53 \pm 0 \cdot 04$	$9 \cdot 56$	$10 \cdot 42$

quantum of one billion years to be added to the numbers in Table 12.2. It seems therefore not out of place to underline the assumptions on which the numbers in Table 12.3 are based.

All of these estimates assume that the earth's lead at one time had the isotopic composition of that now found in iron meteorites. It might seem from inspection of Fig. 12.2 that the reasonable alignment of lead from iron meteorites and chondrites with modern lead strongly supports this assumption. And to some degree this is so, but there are reservations. Thus as more work has been done on the chondrites it has been found that most of them have too little uranium and thorium to account for their radiogenic lead. Hamaguchi *et al.* (1957) showed, by neutron-activation analysis, that these parent concentrations were too low by factors of 2–10. Therefore, while more chondrite points lying close to the 4·55 b.y. isochron are now known than in 1956, most of them have clearly been open systems at some time or other. Modern lead is also not an accurately defined entity. Tilton, who used modern lead ratios with Patterson and Inghram in 1955, discarded them for his age of the earth calculations with Steiger in 1965. Tilton and Steiger (1965) stated:

> Recent papers on the isotopic composition of lead in young volcanic rocks from the ocean basins show that the lead from these rocks is of variable isotopic composition. Values of the age of the earth which vary from 4·42 to 4·65 b.y. can be calculated from these data. Obviously some, if not all the rocks contain lead whose history does not fulfill the three assumptions stated above. [These were essentially the same as assumptions (a) and (b) of Section 12.3.] Close study of the data shows that the closed system requirement has been violated. Variations in the isotopic composition of lead in the basalts show that the uranium–lead ratios in the source materials have differed greatly during the past billion years or so. Possibly other assumptions have been violated as well.

A further puzzle is provided by the observation of radiogenic lead in iron meteorites as well as the very non-radiogenic variety (Starik *et al.*, 1959, 1960; Marshall and Hess, 1961; Murthy, 1964). A single iron meteorite, such as Canyon Diablo, may contain both types of lead in different samples, while the Bischtube meteorite has radiogenic lead in the troilite phase and primordial (non-radiogenic) lead in its metal phase. Murthy (1964) considered that the radiogenic lead was introduced into the irons from chondrites during collisions in space about 500 m.y. ago. Anders (1962) proposed that the radiogenic lead was introduced into the meteorites during terrestrial weathering. The most drastic explanation was offered by Marshall (1962), who suggested that the two sets of lead in iron meteorites were introduced as such at the formation of the meteorites. The problem obviously defies solution at the moment. What is certain, however, from measurement, is that the iron meteorites with the radiogenic lead do not have nearly enough uranium and thorium now to have generated this lead by radioactive decay.

It seems clear from the preceding considerations that we must still view with reserve the assumption that the earth's lead originally had the composition of the non-radiogenic lead component of the troilite phase of iron meteorites. As is often the case, an initially clear, simple picture becomes somewhat blurred as more data are acquired. Tilton and Steiger emphasized that the troilite lead assumption "may be incorrect".

Common to all models is the assumption that the various leads have grown in closed systems since some time very near the earth's formation. As we have already seen in the case of modern lead, this is a requirement more likely to be failed than satisfied. Tilton and Steiger reasoned that errors due to failure of the closed system assumption should be less significant if old leads are used in the calculations, and therefore they chose to concentrate on the lead from Manitouwadge, which is approximately $2 \cdot 7$ b.y. old. Ostic *et al.* (1963) contended that lead from conformable galenas has been generated in a closed system and discounted other leads. That difficulties remain, however, is well illustrated by considerations of the Manitouwadge galena. Tilton and Steiger used merely this galena lead plus the troilitic non-radiogenic lead to calculate the age of the earth. Ostic *et al.* used these same two points and in addition used the isotopic compositions of eight other conformable galenas. Yet the estimates of the earth's age by these two sets of investigators differed by about 200 m.y. Since both groups used troilite lead in the same way, some of the discrepancy must arise from a different outlook with regard to Manitouwadge galena. From a detailed analysis of the lead from felspars and galena, Tilton and Steiger concluded that "the galena data may be taken to characterize the isotopic composition of lead associated with the igneous activity that produced the Algoman granitic rocks at Manitouwadge $2 \cdot 6$ to $2 \cdot 7$ b.y. ago".

Extensive K–Ar, Rb–Sr and U–Pb dating suggested that $2 \cdot 7 \pm 0 \cdot 1$ b.y. ago in fact was the time when the Manitouwadge lead was removed from its uranium- and thorium-bearing parental environment. Ostic *et al.* gave no detailed discussion of this aspect of Manitouwadge lead since their approach, as we have shown earlier, did not involve a knowledge of the age of this galena. However, it is a simple matter to take the values found by these workers for t_0 and V and calculate from their model the age of the Manitouwadge lead. Substituting $t_0 = 4 \cdot 53$ b.y. and $V = 9 \cdot 21$ in equations (12.3) and (12.4) gives $t_{206} = 3 \cdot 02$ b.y. and $t_{207} = 3 \cdot 07$ b.y. Hence, their model requires that the Manitouwadge lead was removed from its parental environment about $3 \cdot 05$ b.y. ago, i.e. approximately 300 m.y. earlier than was estimated by Tilton and Steiger. Here there is a clear divergence of view on the origin of the Manitouwadge lead which affects the calculated age of the earth. That the Manitouwadge area was involved in a great geological hiatus about $2 \cdot 7$ b.y. ago seems beyond doubt from the U–Pb, K–Ar and Rb–Sr analyses. If therefore the lead in the area was extracted from its parental lead-generating environment $3 \cdot 05$ b.y. ago, it must have lain fallow in some non-radioactive environment for 300 m.y. before being incorporated in galena crystals at some stage during the orogeny which took place $2 \cdot 7$ b.y. ago. Whether or not the likelihood of this is remote, it does require the Manitouwadge lead to have undergone at least a two-stage history before its final incorporation into its present environment. Later Ostic *et al.* (1967) concluded that in fact the Manitouwadge lead is the product of at least two environments. Accordingly it should be omitted from the analysis of Ostic *et al.* (1963). When this is done, the value for the age of the earth does not change significantly. However, the basis of the calculation of Tilton and Steiger (1965) is completely removed if the Manitouwadge lead is in fact the product of more than one stage.

12.6. Concordia Method

Ulrych (1967) developed an application of the Wetherill concordia plot (Chapter 5) for the determination of the earth's age. Basalts from the Mid-Atlantic Ridge, the East Pacific Rise, Hawaii, Japan and Easter Island had been analysed by Tatsumoto (1966a, b) for uranium and lead concentrations and for their lead isotope composition. Ulrych represented these data on a condordia plot exactly as is normally done with typical zircon data (Chapter 5). However, instead of subtracting a common lead component of modern-day composition (that is common lead characteristic of the time of effusion of the basalts), which would be pointless as it would essentially remove all the lead, he subtracted lead of the non-radiogenic variety found in iron meteorites. The result was that the basalts from each area yielded points falling about a straight line on the concordia plot as is illustrated for the Hawaiian basalts in Fig. 12.3. The various straight lines all intersected the concordia curve at points we have called t_1 in Chapter 5, where t_1 ranged only from 4·49 to 4·58 b.y. The average value was $t_1 = 4·53 \pm 0·03$ b.y. The basalt lines intersected concordia a second time, at our t_2 of Chapter 5, but with a variety of t_2 values, ranging from 1230 m.y. to 20 m.y. The normal Wetherill interpretation of these data would of course be that the various U–Pb systems considered began life at time t_1 with lead of primeval iron meteoritic composition. The systems remained

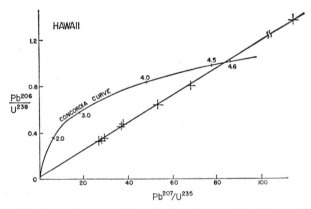

FIG. 12.3. Concordia plot of Hawaiian basalt data. (After Ulrych, 1967.)

closed until the time t_2, when varying amounts of lead and/or uranium were lost or gained. After this episodic disturbance the systems were again closed till recent times. Obviously, however, in this instance the physical picture has to be changed somewhat from the usual zircon case. A grossly oversimplified physical picture for the Hawaiian basalts, for example, would be: uranium–lead systems formed in the mantle at t_1 and remained closed systems for uranium and lead till the time t_2. At this moment the lead was moved, without fractionation or addition of lead of a different composition, into a new environment, presumably the upper mantle, where it was in the presence of a greater amount of uranium than previously. Finally basalt was generated in this environment and poured out on the earth's surface. The mode of formation and eruption of the basalt

was such that the U/Pb ratio of the upper mantle environment was pre-
served in the basalt. Such a simple model fits the observational facts of the
concordia diagram, but it would hardly find favour with petrologists and
geochemists (Oversby and Gast, 1968). Ulrych attributed to the average t_1
the "age of the earth". The t_2 values are difficult to interpret physically and,
unlike the t_1 values, show considerable scatter. The method suffers from
an insensitivity both to significant episodes in the evolutionary history of
the lead isotopes and to large errors in the determination of uranium and
lead concentrations (Oversby and Gast, 1968; Cumming, 1969). It is hard
to avoid the impression that the systematics of this application of the concordia method are not fully understood.

12.7 Conclusion

We have seen that the direct application of U–Pb, Rb–Sr and K–Ar
dating to crustal rocks shows that the earth had at least a small amount of
continental crust 3·5 b.y. ago. Studies of the evolution of lead isotopes, on
the other hand, indicate that this isotopic development has been going on
in the earth for perhaps 4·6 b.y. The question immediately arises therefore
as to why no crustal rock has been dated in the interval 3·5–4·6 b.y.
And this question immediately poses another, which should perhaps be
answered first: to what geophysical event do the lead evolution ages
apply? The lead ages are in fact "model ages" and the models are particularly simple. They all endeavour to calculate when the earth had a uniform
isotopic composition of lead equal to the non-radiogenic variety found in
iron meteorites. With what geophysical event this is coincident is a matter
for speculation. It might represent the time when the earth grew by
accretion from cold planetesimals of meteoritic composition, never to
become molten; or it might date a hot, completely molten beginning for
the earth. Regardless, however, of our uncertainty regarding the physical
nature of the event being dated, the evidence presented in this chapter
does seem to indicate that the terrestrial and meteoritic lead began to
diverge from a common isotopic composition approximately 4·6 b.y. ago,
and we may currently take this number to be "the age of the earth"
within a few hundred million years. Confidence in this value is increased
by the fact that K–Ar and Rb–Sr ages of meteorites independently indicate
that the meteorites underwent some severe physical-chemical disturbance
4·6 b.y. ago. However, with the present body of experimental knowledge
it is not yet possible to explain the one billion year difference between the

age of the oldest rocks and the ages found from lead evolution studies. If the earth formed at a relatively low temperature, it may be that it required a billion years of radioactive heating to become active enough to build continents as we conceive of them. But as Holmes (1965) said "... here we are in the shadows of speculation and must await the illumination of further discoveries".

REFERENCES

ADAMS, J. A. S., EDWARDS, G., HERLE, W. and OSMOND, K. (1958) *Bull. Geol. Soc. Amer.* **69**, 1527.
AHRENS, L. H. (1955) *Geochim. Cosmochim. Acta* **7**, 294.
ALDRICH, L. T. and NIER, A. O. (1948) *Phys. Rev.* **74**, 876.
ALDRICH, L. T., DAVIS, G. L. and JAMES, H. L. (1965) *J. Petrol.* **6**, 445.
ALDRICH, L. T. and WETHERILL, G. W. (1958) *Ann. Rev. Nucl. Sci.* **8**, 257.
ALDRICH, L. T., WETHERILL, G. W., DAVIS, G. L. and TILTON, G. R. (1958) *Trans. Amer. Geophys. Union* **39**, 1124.
ALDRICH, L. T., WETHERILL, G. W., TILTON, G. R. and DAVIS, G. L. (1956) *Phys. Rev.* **103**, 1045.
ALLSOPP, H. L. (1961) *J. Geophys. Res.* **66**, 1499.
ALLSOPP, H. L., ROBERTS, H. R., SCHREINER, G. D. L. and HUNTER, D. R. (1962) *J. Geophys. Res.* **67**, 5307.
ALPHER, R. A., BETHE, H. A. and GAMOW, G. (1948) *Phys. Rev.* **73**, 803.
ALPHER, R. A. and HERMAN, R. C. (1951) *Phys. Rev.* **84**, 1111.
AMARAL, G., CORDANI, U. G., KAWASHITA, K. and REYNOLDS, J. H. (1966) *Geochim. Cosmochim. Acta* **30**, 159.
AMIRKHANOFF, K. L., BRANDT, S. B. and BARNITSKY, E. N. (1961) *Ann. N.Y. Acad. Sci.* **91**, 2.
ANDERS, E., (1962) *Rev. Mod. Phys.* **34**, 287.
ANDERS, E. (1963) *The Moon, Meteorites and Comets, The Solar System*, Vol. 4, Univ. Chicago Press.
ANDERS, E. (1964) *Space Sci. Rev.* **3**, 583.
ARMSTRONG, R. L. (1966a) *Geochim. Cosmochim. Acta* **30**, 565.
ARMSTRONG, R. L. (1966b) in *Potassium Argon Dating*, Springer, New York, p. 117.
ARMSTRONG, R. L., JÄGER, E. and EBERHARDT, P. (1966) *Earth Planet. Sci. Letters* **1**, 13.
ASTON, F. W. (1933) *Proc. Roy. Soc.* A **140**, 535.

BAADSGAARD, H. and DODSON, M. H. (1964) *Quart. J. Geol. Soc. Lond.* **120 S**, 119.
BAADSGAARD, H., LIPSON, J. and FOLINSBEE, R. E. (1961) *Geochim. Cosmochim. Acta* **25**, 147.
BAILEY, S. W., HURLEY, P. M., FAIRBAIRN, H. W. and PINSON, W. H., Jr. (1962) *Bull. Geol. Soc. Amer.* **73**, 1167.
BAKSI, A. K., YORK, D. and WATKINS, N. D. (1967) *J. Geophys. Res.* **72**, 6299.
BARNARD, G. P. (1952) *Modern Mass Spectrometry*, The Institute of Physics, London.
BARRER, R. M. and FALCONER, J. D. (1956) *Proc. Roy. Soc.* A **236**, 227.
BOFINGER, V. M. and COMPSTON, W. (1967) *Geochim. Cosmochim. Acta* **31**, 2353.
BOGARD, D. D., BURNETT, D. S., EBERHARDT, P. and WASSERBURG, G. J. (1967) *Earth Planet. Sci. Letters* **3**, 179.
BONHOMMET, N. and BABKINE, J. (1967) *Compt. Rend.* **264**, 92.
BONHOMMET, N. and ZÄHRINGER, J. (1969) *Earth Planet. Sci. Letters* **6**, 43.
BROWN, H., (1947) *Phys. Rev.* **72**, 348.

BRUCKSHAW, J. M. and ROBERTSON, E.I. (1949) *Mon. Not. Roy. Astr. Soc. Geophys. Suppl.* **5**, 308.

BRUNHES, B. (1906) *J. Physique*, **5**, 705.

BULLARD, E. C. (1949) *Proc. Roy. Soc. A* **199**, 413.

BULLARD, E. C. and STANLEY, J. P. (1949) *Publ. Finn. Geodet. Inst.* **36**, 33.

BURNETT, D. S., LIPPOLT, H. J. and WASSERBURG, G. J. (1966) *J. Geophys. Res.* **71**, 1249.

BURNETT, D. S. and WASSERBURG, G. J. (1967a) *Earth Planet. Sci. Letters* **2**, 137.

BURNETT, D. S. and WASSERBURG, G. J. (1967b) *Earth Planet. Sci. Letters* **2**, 397.

BYSTRÖM-ASKLUND, A. M., BAADSGAARD, H. and FOLINSBEE, R. E. (1961) *Geol. Fören. Stockh. Förh.* **83**, 92.

CAMPBELL, C. D. and RUNCORN, S. K. (1956) *J. Geophys. Res.* **61**, 449.

CAMPBELL, N. R. and WOOD, A. (1906) *Proc. Camb. Phil. Soc.* **14**, S, 15.

CATANZARO, E. J. (1963) *J. Geophys. Res.* **68**, 2045.

CATANZARO, E. J. (1967) *J. Geophys. Res.* **72**, 1325.

CATANZARO, E. J. and KULP, J. L. (1964) *Geochim. Cosmochim. Acta* **28**, 87.

CHANDRASEKHAR, S. (1953) *Phil. Mag.* **44**, 233, 1129.

CHEVALLIER, R. (1925) *Ann. Phys. Paris*, ser. 10, **4**, 5.

CHOW, T. J. and PATTERSON, C. C. (1962) *Geochim. Cosmochim. Acta* **26**, 263.

CLARK, F. L., SPENCER-PALMER, H. J. and WOODWARD, R. N. (1944) Bull. Jealott's Hill Research Sta., Imp. Chem. Ind., Ltd.; Declassified Rept. BR-522 (unpublished data, 1944).

COBB, J. C. and KULP, J. L. (1960) *Bull. Geol. Soc. Amer.* **71**, 223.

COBB, J. C. and KULP, J. L. (1961) *Geochim. Cosmochim. Acta* **24**, 226.

COLLINS, C. B., RUSSELL, R. D. and FARQUHAR, R. M. (1953) *Can. J. Phys.* **31**, 402.

COMPSTON, W. and JEFFERY, P. M. (1959) *Nature* **184**, 1792.

COMPSTON, W., LOVERING, J. F. and VERNON, M. J. (1965) *Geochim. Cosmochim. Acta* **29**, 1085.

COOPER, J. A. (1963) *Geochim. Cosmochim. Acta* **27**, 525.

COOPER, J. A. and RICHARDS, J. R. (1966) *Earth Planet. Sci. Letters* **1**, 58.

CORMIER, R. F. (1956) *Bull. Geol. Soc. Amer.* (Abstract) **67**, 1812.

CORMIER, R. F. and KELLY, A. M. (1964) *Can. J. Earth Sci.* **1**, 159.

COX, A. (1969) *Science* **163**, 237.

COX, A. and DALRYMPLE, G. B. (1967) *J. Geophys. Res.* **72**, 2603.

COX, A., DOELL, R. R. and DALRYMPLE, G. B. (1963a) *Science* **142**, 382.

COX, A., DOELL, R. R. and DALRYMPLE, G. B. (1963b) *Nature* **198**, 1049.

COX, A., DOELL, R. R. and DALRYMPLE, G. B. (1964) *Science* **144**, 1539.

CROUCH, E. A. C. and WEBSTER, R. K. (1963) *J. Chem. Soc.* **18**, 118.

CUMMING, G. L. (1969) *Can. J. Earth Sci.* **6**, 719.

CURRY, R. R. (1966) *Science* **154**, 770.

CURTIS, L. F., STOCKMAN, L. and BROWN, B.W. (1941) Natl. Bur. Standards (U.S.), Rept. A80 (unpublished data, 1941); discussed in Fleming *et al.* (1952).

DALRYMPLE, G. B. (1964) *Bull. Geol. Soc. Amer.* **75**, 753.

DALRYMPLE, G. B. (1967) *Earth Planet. Sci. Letters* **3**, 289.

DALRYMPLE, G. B. and HIROOKA, K. (1965) *J. Geophys. Res.* **70**, 5291.

DAMON, P. E. and KULP, J. L. (1957) *Amer. Min.* **43**, 433.

DAVIS, G. L. and TILTON, G. R. (1959) *Researches in Geochemistry*, ed. P. H. ABELSON.

DEARNLEY, R. (1965) *Nature* **206**, 1083.

DEER, W. A., HOWIE, R. A. and ZUSSMAN, J. (1964) *Rock-forming Minerals*, Longmans.

DICKE, R. H. (1959) *Nature* **183**, 170.

DIETZ, R. S. (1961) *Nature* **190**, 854.
DIRAC, P. A. M. (1939) *Proc. Roy. Soc.* A **165**, 199.
DODSON, M. H. (1963a) D.Phil. Thesis, Oxford.
DODSON, M. H. (1963b) *J. Sci. Instr.* **40**, 289.
DODSON, M. H., REX, D. C., CASEY, R. and ALLEN, P. (1964) *Quart. J. Geol. Soc. Lond.* **120S**, 145.
DOE, B. R. (1962) *J. Geophys. Res.* **67**, 2895.
DOELL, R. R., DALRYMPLE, G. B. and COX, A. (1966) *J. Geophys. Res.* **71**, 531.
DUCKWORTH, H. E. (1958) *Mass Spectroscopy*, Cambridge University Press.

ELSASSER, W. M. (1939) *Phys. Rev.* **55**, 489.
ENGEL, A. J. (1963) *Science* **140**, 143.
ERICSON, D. B., EWING, M. and WOLLIN, G. (1964) *Science* **146**, 723.
ERICKSON, G. P. and KULP, J. L. (1961) *Bull. Geol. Soc. Amer.* **72**, 649.
EVERNDEN, J. F. and CURTIS, G. H. (1965) *Current Anthropology* **6**, 343.
EVERNDEN, J. F., CURTIS, G. H., OBRADOVICH, J. and KISTLER R. W. (1961) *Geochim. Cosmochim. Acta* **23**, 78.
EVERNDEN, J. F., CURTIS, G. H., SAVAGE, D. E. and JAMES, G. T. (1964) *Amer. J. Sci.* **262**, 145.
EVERNDEN, J. F. and JAMES, G. T. (1964) *Amer. J. Sci.* **262**, 945.
EVERNDEN, J. F. and RICHARDS, J. R. (1962) *J. Geol. Soc. Aust.* **9**, 1.

FAHRIG, W. F. and WANLESS, R. K. (1963) *Nature* **200**, 934.
FARRAR, E., McINTYRE, R. M., YORK, D. and KENYON, W. J. (1964) *Nature* **204**, 531.
FAUL, H. (1960) *Bull. Geol. Soc. Amer.* **71**, 637.
FAUL, H. and THOMAS, H. (1959) (Abstract), *Bull. Geol. Soc. Amer.* **70**, 1600.
FECHTIG, H. and KALBITZER, S. (1966) in *Potassium Argon Dating*, Springer, New York, p. 68.
FECHTIG, H., GENTNER, W. and KALBITZER, S. (1960) *Geochim. Cosmochim. Acta* **19**, 70.
FECHTIG, H., GENTNER, W. and KALBITZER, S. (1961) *Geochim. Cosmochim. Acta* **25**, 297.
FISH, R. A. and GOLES, G. G. (1962) *Nature* **196**, 27.
FISHER, D. E. (1965) *J. Geophys. Res.* **70**, 2445.
FITCH, F. J. and MILLER, J. A. (1965) *Nature* **206**, 1023.
FLEISCHER, R. L. and PRICE, P. B. (1964) *Geochim. Cosmochim. Acta* **28**, 1705.
FLEISCHER, R. L., PRICE, P. B. and WALKER, R. M. (1964) *J. Geophys. Res.* **69**, 4885.
FLEISCHER, R. L., PRICE, P. B. and WALKER, R. M. (1965) *J. Geophys. Res.* **70**, 1497.
FLEMING, E. H., GHIORSO, A. and CUNNINGHAM, B. B. (1952) *Phys. Rev.* **88**, 642.
FLINT, R. F. (1965) *Geol. Soc. Amer. Spec. Paper* **84**, 497.
FLYNN, K. F. and GLENDENNIN, L. E. (1959) *Phys. Rev.* **116**, 744.
FOLINSBEE, R. E., BAADSGAARD, H. and LIPSON, J. (1960) *Int. Geol. Congr.* **21** (3), 7.

GABRIELSE, H., and REESOR, J. (1964) *Roy. Soc. Can. Spec. Publ.* No. 8, 96.
GAST, P. W. (1960) *J. Geophys. Res.* **65**, 1287.
GAST, P. W. (1962) *Geochim. Cosmochim. Acta* **26**, 927.
GASTIL, G. (1960) *Amer. J. Sci.* **258**, 1.
GENTNER, W., GOEBEL, K. and PRÄG, R. (1954) *Geochim. Cosmochim. Acta* **5**, 124.
GENTNER, W., PRÄG, R. and SMITS, F. (1953) *Geochim. Cosmochim. Acta* **4**, 11.
GERLING, E. K. (1942) *Dokl. Akad. Nauk SSSR*, **34**, 259.
GERLING, E. K. and MOROZOVA, I. M. (1957) *Geochemistry* No. 4.
GERLING, E. K. and PAVLOVA, T. G. (1951) *Doklady Akad. Nauk SSSR* **77**, 85.

GERLING, E. K., SHUKOLYUKOV, YU. A., KOL'TSOVA, T. V., MATVEYEVA, I. I. and YAKOVLEVA, S. Z. (1962) *Geochemistry* 1055.

GILETTI, B. J., and KULP, J. L. (1955) *Amer. Mineral.* **40,** 481.

GILLULY, J. (1949) *Bull. Geol. Soc. Amer.* **60,** 561.

GITTINS, J., MACINTYRE, R. M. and YORK, D. (1967) *Can. J. Earth Sci.* **4,** 651.

GOLDICH, S. S. and HEDGE, C. E. (1962) *J. Geophys. Res.* **67,** 3561.

GOLDICH, S. S., BAADSGAARD, H. and NIER, A. O. (1957) *Trans. Amer. Geophys. Union* **38,** 547.

GOLDICH, S. S., NIER, A. O., BAADSGAARD, H., HOFFMAN, J. H. and KRUEGER, H. W. (1961) *Minnesota Geol. Surv. Bull.* **41.**

GOLES, G. G. and ANDERS, E. (1962) *Geochim. Cosmochim. Acta* **26,** 723.

GRANT, J. A. (1964) *Science* **146,** 1049.

GRASTY, R. L. and MILLER, J. A. (1965) *Nature* **207,** 1146.

HAHN-WEINHEIMER, P. and ACKERMANN, H. (1963) *Z. Analyt. Chem.* **194,** 81.

HALLER, J. and KULP, J. L. (1962) *Medd. Gronland* **171** (1), 1.

HAMAGUCHI, H., REED, G. W. and TURKEVICH, A. (1957) *Geochim. Cosmochim. Acta* **12,** 337.

HAMILTON, E. I. (1965) *Applied Geochronology,* Academic Press, London and New York.

HAMILTON, E. I., DODSON, M. H. and SNELLING, N. J. (1962) *Int. J. Appl. Radiation and Isotopes* **13,** 587.

HARPER, C. T. (1964) *Nature* **203,** 468.

HARPER, C. T. (1967) *Earth Planet. Sci. Letters* **3,** 128.

HARRIS, P. M., FARRAR, E., McINTYRE, R. M., YORK, D. and MILLER, J. A (1965) *Nature* **205,** 352.

HART, S. R. (1961) *J. Geophys. Res.* **66,** 2995.

HART, S. R. (1964) *J. Geol.* **72,** 493.

HART, S. R. (1966) *Trans. Amer. Geophys. Union* **47,** 280.

HART, S. R. and DODD, R. J. (1962) *J. Geophys. Res.* **67,** 2998.

HERZOG, L. F. and PINSON, W. H., Jr. (1956) *Amer. J. Sci.* **254,** 555.

HERZOG, L. F., PINSON, W. H. and CORMIER, R. F. (1958) *Bull. Amer. Ass. Petrol. Geol.* **42,** 717.

HESS, H. H. (1962) *Geol. Soc. Amer. Buddington,* Vol. 599.

HIDE, R. (1967) *Science* **157,** 55.

HIRT, B., TILTON, G. R., HERR, W. and HOFFMEISTER, W. (1963) in *Earth Sciences and Meteorites,* eds. J. GEISS and E. D. GOLDBERG, North-Holland, Amsterdam.

HOHENBERG, C. M., PODOSEK, F. A. and REYNOLDS, J. H. (1967) *Science* **156,** 233.

HOLMES, A. (1928–9) *Trans. Geol. Soc. Glasgow* **18,** 579.

HOLMES, A. (1931) *Bull. Nat. Res. Coun., Wash.* **80,** 124.

HOLMES, A. (1946) *Nature* **157,** 680.

HOLMES, A. (1947) *Nature* **159,** 127.

HOLMES, A. (1949) *Nature* **163,** 453.

HOLMES, A. (1959) *Trans. Edinb. Geol. Soc.* **17,** 183.

HOLMES, A. (1965) *Principles of Physical Geology,* London: Nelson.

HOSPERS, J. (1953–4) Kon. Ned. Akad. Wet. Proc. of the section of sciences, Series B, *Phys. Sci.* Part I: 56, 467, 1953; Part II: 56, 477, 1953; Part III: 57, 112, 1954.

HOUTERMANS, F. (1946) *Naturwiss.* **33,** 183.

HOUTERMANS, F. (1947) *Z. Naturforsch.* **2a,** 322.

HOWER, J., HURLEY, P. M., PINSON, W. H. and FAIRBAIRN, H. W. (1963) *Geochim. Cosmochim. Acta* **27,** 405.

HUNTER, D. R. (1957) *Trans. Geol. Soc. S. Africa* **60**, 85.

HURLEY, P. M. (1958) *5th Ann. Prog. Rept. Mass. Inst. Tech.*, pp. 1–3.

HURLEY, P. M. (1966) *Potassium Argon Dating*, Springer, New York, p. 134.

HURLEY, P. M., BOUCOT, A. J., ABEE, A. L., FAUL, H., PINSON, W. H. and FAIRBAIRN H. W. (1959) *Bull. Geol. Soc. Amer.* **70**, 947.

HURLEY, P. M., CORMIER, R. F., HOWER, J., FAIRBAIRN, H. W. and PINSON, W. H. (1960) *Bull. Amer. Ass. Petrol. Geol.* **44**, 1793.

HURLEY, P. M., HUGHES, H., FAURE, G., FAIRBAIRN, H. W. and PINSON, W. H. (1962) *J. Geophys. Res.* **67**, 5315.

HURLEY, P. M., HUGHES, H., PINSON, W. H., Jr. and FAIRBAIRN, H. W. (1963) *Geochim. Cosmochim. Acta* **26**, 67.

INGHRAM, M. G., BROWN, H., PATTERSON, C. C. and HESS, D. C. (1950) *Phys. Rev.* **80**, 916.

INGHRAM, M. G. and HAYDEN, R. J. (1954) *Handbook on Mass Spectroscopy*, Nucl. Sci. Series Rept., 14, Nat. Res. Council.

IRVING, E. (1964) *Palaeomagnetism and its Application to Geological and Geophysical Problems*, Wiley, New York.

IRVING, E. (1967) *J. Geophys. Res.* **71**, 6025.

JÄGER, E. (1965) in *Géochronologie Absolue*, Centre National de la Recherche Scientifique, No. 151, Nancy, 3–8 Mai, p. 191.

JEFFERY, P. M. and REYNOLDS, J. H. (1961) *Z. Naturforsch.* **16a**, 431.

JOLY, J. (1899) *Sci. Trans. R. Dublin Soc.* **7**, 23.

KANASEWICH, E. R. (1962) *Geophys. J.* **7**, 158.

KANASEWICH, E. R. and SAVAGE, J. C. (1963) *Can. J. Phys.* **41**, 1911.

KAVILADZE, M. S. and ABASHIDZE, I. V. (1964) *Bull. Acad. Sci. Georgian SSR* **35**, 67.

KAZAKOV, G. A. and POLEVAYA, N. I. (1958) *Geochemistry* **1958** (4), 374.

KELVIN, LORD (1899) *Journ. Victoria Ins. London* **331**, 11.

KENDALL, B. R. F. (1960) *Nature* **186**, 225.

KENDALL, M. G. and SMITH, B. B. (1939) *Tables of Random Sampling Numbers*, Cambridge University Press, London.

KHLOPIN, V. G. and ABIDOV, S. H. (1941) *Dokl. Akad. Nauk SSR* **32**, 637.

KIENBERGER, C. A. (1949) *Phys. Rev.* **76**, 1561.

KNIGHT, G. B. (1950) U.S. Atomic Energy Commission Report, O.R.N.L. K-663 (unpublished data, 1950).

KOLLAR, F., RUSSELL, R. D. and ULRYCH, T. J. (1960) *Nature* **187**, 754.

KOVARIK, A. F., and ADAMS, N. I., Jr. (1938) *Phys. Rev.* **54**, 413.

KOVARIK, A. F. and ADAMS, N. I., Jr. (1941) *J. Appl. Phys.* **12**, 296.

KOVARIK, A. F. and ADAMS, N. I., Jr. (1955) *Phys. Rev.* **98**, 46.

KRAUSHAAR, J. J., WILSON, E. D. and BAINBRIDGE, K. T. (1953) *Phys. Rev.* **90**, 510.

KRYLOV, A. Y. (1961) *Ann. N.Y. Acad. Sci.* **91**, 324.

KULP, J. L. (1961) *Science* **133**, 1105.

KULP, J. L. and ENGELS, J. (1963) *Proc. Symp. Radioactive Dating*, Athens, 19–23 Nov. 1962, Intern. At. Energy Agency, STI, Publ. 68, Vienna, 219.

LAMBERT, R. St. J. (1964) *Quart. J. Geol. Soc. Lond.* **120S**, 43.

LANPHERE, M. A. and DALRYMPLE, G. B. (1966) *Nature* **209**, 902.

LANPHERE, M. A., WASSERBURG, G. J., ALBEE, A. L. and TILTON, G. R. (1964) in *Isotopic and Cosmic Chemistry*, North-Holland, Amsterdam, p. 269.

LEDENT, D., LAY, C. and DELHAL, J. (1962) *Bull. Soc. Belg. Geol.* **71**, 223.
LEECH, A. P. (1966) *Can. J. Earth Sci.* **3**, 389.
LEININGER, R. F., SEGRÉ, E. and WIEGAND, C. (1951) *Phys. Rev.* **81**, 280.
LÉTOLLE, R. (1963) *Compt. Rend.* **257**, 3996.
LEUTZ, H., WENNINGER, H. and ZIEGLER, K. (1962) *Z. Phys.* **169**, 409.
LIPPOLT, H. J. and GENTNER, W. (1963) in *Radioactive Dating*, p. 239, I.A.E.A., Vienna.
LIPSON, J. I. (1956) *Geochim. Cosmochim. Acta* **10**, 149.
LIVINGSTON, D. E., DAMON, P. E., MAUGER, R. L., BENNETT, R. and LAUGHLIN, A. W. (1967) *J. Geophys. Res.* **72**, 1361.
LIVINGSTON, D. (1963) *Geochim. Cosmochim. Acta* **27**, 1055.
LONG, L. E. (1966) *Earth Planet Sci. Letters* **1**, 289.
LOVERING, J. F. and RICHARDS, J. R. (1964) *J. Geophys. Res.* **69**, 4895.

MARSHALL, R. R. (1962) *Icarus* **1**, 95.
MARSHALL, R. R. and HESS, D. C. (1960) *Analyt. Chem.* **32**, 960.
MARSHALL, R. R. and HESS, D. C. (1961) *Geochim. Cosmochim. Acta* **21**, 161.
MASON, B. (1962) *Meteorites*, Wiley, New York.
MASON, R. G. (1958) *Geophys. J.* **1**, 320.
MATUYAMA, M. (1929) *Proc. 4th Pacific Science Congress* **1**.
MAYNE, K. I. (1952) *Rep. Progr. Physics* **15**, 24.
MCCARTNEY, W. D., POOLE, W. H., WANLESS, R. K., WILLIAMS, H. and LOVERIDGE, W. D. (1966) *Can. J. Earth Sci.* **3**, 947.
MCDOUGALL, I. (1961) *Nature* **190**, 1184.
MCDOUGALL, I. (1963) *J. Geophys. Res.* **68**, 1535.
MCDOUGALL, I. (1966) in *Methods and Techniques in Geophysics*, Interscience, New York, p. 279.
MCDOUGALL, I. and CHAMALAUN, J. H. (1966) *Nature* **212**, 1415.
MCDOUGALL, I., COMPSTON, W. and BOFINGER, V. M. (1966) *Bull. Geol. Soc. Amer.* **77**, 1075.
MCDOUGALL, I., COMPSTON, W. and HAWKES, D. D. (1963) *Nature* **198**, 564.
MCDOUGALL, I. and GREEN, D. H. (1964) *Norsk. Geol. Tidskr.* **44**, pt. 2, 183.
MCDOUGALL, I. and RÜEGG, N. R. (1966) *Geochim. Cosmochim. Acta* **30**, 191.
MCDOUGALL, I. and STIPP, J. J. (1969) *Trans. Amer. Geophys. Union* **50**, 330.
MCDOUGALL, I. and TARLING, D. H. (1963) *Nature* **200**, 54.
MCDOUGALL, I. and WENSINK, H. (1966) *Earth Planet. Sci. Letters* **1**, 232.
MCINTYRE, G. A., BROOKS, C., COMPSTON, W. and TUREK, A. (1966) *J. Geophys. Res.* **71**, 5459.
MCINTYRE, R. M., YORK, D. and GITTINS, J. (1966) *Nature* **209**, 702.
MCINTYRE, R. M., YORK, D. and MOORHOUSE, W. W. (1967) *Can. J. Earth Sci.* **4**, 815.
MCMULLEN, C. C., FRITZE, K. and TOMLINSON, R. H. (1966) *Can. J. Phys.* **44**, 3033.
MCNAIR, A., GLOVER, R. N. and WILSON, H. W. (1956) *Phil. Mag.* **1**, 199.
MERCANTON, P. (1926) *C.R. Acad. Sci. Paris* **151**, 1092.
MERRIHUE, C. M. (1965) *Trans. Amer. Geophys. Union* **46**, 125.
MERRIHUE, C. M. and TURNER, G. (1966) *J. Geophys. Res.* **71**, 2852.
MILLER, J. A. and FITCH, F. J. (1964) *Quart. J. Geol. Soc. Lond.* **120S**, 101.
MOORBATH, S. (1965) in *Controls of Metamorphism*, Oliver & Boyd, Edinburgh.
MOORBATH, S. and BELL, J. D. (1965) *J. Petrol.* **6**, 37.
MÜLLER, O. and ZÄHRINGER, J. (1966) *Geochim. Cosmochim. Acta* **30**, 1075.
MURTHY, V. R. (1964) *Isotopic and Cosmic Chemistry*, North-Holland, Amsterdam.
MURTHY, V. R. and COMPSTON, W. (1965) *J. Geophys. Res.* **70**, 5297.
MURTHY, V. R. and PATTERSON, C. C. (1962) *J. Geophys. Res.* **67**, 1161.

NAGATA, T. (1961) *Rock Magnetism*, Maruzen, Tokyo.
NEUVONEN, K. J. (1961) *Bull. Comm. Geol. Finlande* **196**, 445.
NICOLAYSEN, L. O. (1957) *Geochim. Cosmochim. Acta* **11**, 41.
NICOLAYSEN, L. O. (1961) *Ann. N.Y. Acad. Sci.* **91**, 198.
NICOLAYSEN, L. O., BURGER, A. J. and LIEBENBERG, W. R. (1962) *Geochim. Cosmochim. Acta* **26**, 15.
NIER, A. O. (1938) *J. Amer. Chem. Soc.* **60**, 1571.
NIER, A. O. (1939) *Phys. Rev.* **55**, 153.
NIER, A. O. (1950a) *Phys. Rev.* **77**, 789.
NIER, A. O. (1950b) *Phys. Rev.* **79**, 450.
NIER, A. O., THOMSON, R. W. and MURPHY, B. F. (1941) *Phys. Rev.* **60**, 112.

OPDYKE, N. D., GLASS, B., HAYS, J. D. and FOSTER, J. (1966) *Science* **154**, 349.
OSTIC, R. G., RUSSELL, R. D. and REYNOLDS, P. H. (1963) *Nature* **199**, 1150.
OVERSBY, V. M. and GAST, P. W. (1968) *Science* **162**, 925.

PALMER, G. H. and AITKEN, K. L. (1953) *J. Sci. Inst.* **30**, 314.
PATTERSON, C. C. (1951) Atomic Energy Commission Report, A.E.C.D.-3180, 1951.
PATTERSON, C. C. (1953) *Proc. Conf. Nuc. Processes in Geol. Settings, Williams Bay*, Sept. 1953, pp. 36–40.
PATTERSON, C. C. (1955) *Nuclear Processes in Geological Settings* (Penn. State University; Nat. Res. Council, Pub. 400), p. 157.
PATTERSON, C. C. (1956) *Geochim. Cosmochim. Acta* **10**, 230.
PATTERSON, C. C. (1964) *Isotopic and Cosmic Chemistry*, North-Holland, Amsterdam, p. 244.
PATTERSON, C. C. and TATSUMOTO, M. (1964) *Geochim. Cosmochim. Acta* **28**, 1.
PATTERSON, C. C., TILTON, G. R. and INGHRAM, M. G. (1955) *Science* **121**, 69.
PHILPOTTS, A. R. and MILLER, J. A. (1963) *Geol. Mag.* **100**, 337.
PICCIOTTO, E. and WILGAIN, S. (1956) *Nuovo Cimento* **4**, 1525.
PINSON, W. H., FAIRBAIRN, H. W. and CORMIER, R. F. (1958) *Bull. Geol. Soc. Amer.* **69**, 599.
PINSON, W. H., SCHNETZLER, C. C., BEISER, E., FAIRBAIRN, H. W. and HURLEY, P. M. (1965) *Geochim. Cosmochim. Acta* **29**, 455.
PITMANN, W. C. and HEIRTZLER, J. R. (1966) *Science* 1170.
POLEVAYA, N. I., TITOV, H. E., BELYAEV, V. A. and SPRINTSSON, V. D. (1958) *Geochemistry* **1958** (8), 897.
POLKANOV, A. A. and GERLING, E. K. (1961) *Ann. N.Y. Acad. Sci.* **91**, 492.
PREY, A. (1922) *Abh. Ges. der Wissenschaften Math. Phys. Klasse (Gottingen)* **11**, 1.
PRICE, P. B. and WALKER, R. M. (1962) *Nature* **196**, 732.

RAMA, S. N. I., HART, S. R. and ROEDDER, E. (1965) *J. Geophys. Res.* **70**, 509.
RANCITELLI, L., FISHER, D. E., FUNKHOUSER, J. and SCHAEFFER, D. A. (1967) *Science* **155**, 999.
RANKAMA, K. (1954) *Isotope Geology*, Interscience, N.Y.
RANKAMA, K. (1963) *Progress in Isotope Geology*, Interscience, N.Y.
RAYLEIGH, LORD (1933) *Proc. Roy. Soc.* A **142**, 370.
REED, G. W. and TURKEVICH, A. (1955) *Nature* **176**, 794.
REED, G. W., HAMAGUCHI, H. and TURKEVICH, A. (1958) *Geochim. Cosmochim. Acta* **13**, 248.
REUTERSWÄRD, C. (1952) *Ark. Fys.* **4**, 203.
REUTERSWÄRD, C. (1956) *Ark. Fys.* **11**, 1.

REYNOLDS, J. H. (1956) *Rev. Sci. Inst.* **27**, 928.
REYNOLDS, J. H. (1957) *Geochim. Cosmochim. Acta* **12**, 177.
REYNOLDS, J. H. (1960) *Phys. Rev. Letters* **4**, 8.
REYNOLDS, J. H. (1963) *J. Geophys. Res.* **68**, 2939.
REYNOLDS, J. H. and TURNER, G. (1964) *J. Geophys. Res.* **69**, 3263.
ROCHE, A. (1953) *Compt. Rend. Acad. Sci.* **236**, 107.
ROSS, J. V. (1966) *Can. J. Earth Sci.* **3**, 259.
RUNCORN, S. K. (1962) *Nature* **193**, 311.
RUSSELL, H. N. (1921) *Proc. Roy. Soc.* A **99**, 84.
RUSSELL, R. D. (1963) in *Earth Sciences and Meteoritics*, North-Holland, Amsterdam, p. 44.
RUSSELL, R. D. and AHRENS, L. H. (1957) *Geochim. Cosmochim. Acta* **11**, 213.
RUSSELL, R. D. and FARQUHAR, R. M. (1960) *Lead Isotopes in Geology*, Interscience, New York.
RUTHERFORD, E. (1900) *Phil. Mag.*, 5th Series, **49**, 1.

SARDAROV, S. S. (1957) *Geochemistry* **3**, 193.
SARDAROV, S. S. (1963) *Geochemistry* **10**, 937.
SAYAG, G. J. (1951) *Compt. Rend.* **232**, 2091.
SCHAEFFER, O. A. and ZÄHRINGER, J. (1966) *Potassium Argon Dating*, Springer, New York.
SCHREINER, G. D. L. (1958) *Proc. Roy. Soc.* A **245**, 112.
SCHREINER, G. D. L. and VERBEEK, A. A. (1965) *Proc. Roy. Soc.* A **285**, 423.
SCHREYER, W., YODER, H. S. and ALDRICH, L. T. (1960) *Ann. Rept. Geophys. Lab. Carnegie Inst. Washington, Year Book* **59**, 91.
SCHUMACHER, E. (1956) *Z. Naturforsch.* **11a**, 206.
SENFTLE, F. E., FARLEY, T. A. and LAZAR, N. (1956) *Phys. Rev.* **104**, 1629.
SHIELDS, W. R., GARNER, E. L., HEDGE, C. E. and GOLDICH, S. S. (1963) *J. Geophys. Res.* **68**, 2331.
SHIELDS, P. M., PINSON, W. H., Jr. and HURLEY, P. M. (1966) *J. Geophys. Res.* **71**, 2163.
SHIMA, M. and HONDA, M. (1967) *Earth Planet. Sci. Letters* **2**, 344.
SHOEMAKER, E. M. (1963) *The Moon Meteorites and Comets*, eds. B. M. MIDDLEHURST and G. P. KUIPER, Univ. of Chicago Press.
SILK, E. C. H. and BARNES, R. S. (1959) *Phil. Mag.* **4**, 970.
SILVER, L. T. and DEUTSCH, S. (1963) *J. Geol.*, **71**, 721.
DE SITTER, L. U. (1959) *Structural Geology*, McGraw-Hill, New York.
SMITH, A. G. (1964) *Quart. J. Geol. Soc. Lond.* **120S**, 129.
SMITH, J. D. and FOSTER, J. H. (1969) *Science* **163**, 565.
SMITH, J. V. and SCHREYER, W. (1962) *Min. Mag.* **33**, 226.
SMITS, F. and GENTNER, W. (1950) *Geochim. Cosmochim. Acta* **1**, 22.
STANTON, R. and RUSSELL, R. D. (1959) *Econ. Geol.* **54**, 588.
STARIK, I. E., SOBOTOVICH, E. V., LOVTSYUS, G. P., SHATS, M. M. and LOVTSYUS, A. V. (1959) *Dokl. Akad. Nauk SSSR* **123**, 688.
STARIK, I. E., SOBOTOVICH, E. V., LOVTSYUS, G. P., SHATS, M. M. and LOVTSYUS, A. V. (1960) *Dokl. Akad. Nauk SSSR* **134**, 555.
STEIGER, R. (1964) *J. Geophys. Res.* **69**, 5407.
STIPP, J. J., CHAPPELL, J. M. A. and McDOUGALL, I. (1967) *Amer. J. Sci.* **265**, 462.
STOENNER, R. W. and ZÄHRINGER, J. (1958) *Geochim. Cosmochim. Acta* **15**, 40.
STORZER, D. and WAGNER, G. A. (1969) *Earth Planet. Sci. Letters* **5**, 463.
STRUTT, R. J. (1908) *Proc. Roy. Soc.* A **80**, 572.
SUTTON, J. (1963) *Nature* **198**, 731.

TATSUMOTO, M. (1966a) *Science* **153**, 1094.
TATSUMOTO, M. (1966b) *J. Geophys. Res.* **71**, 1721.
THOMSON, J. J. (1905) *Phil. Mag.* **10**, 584.
TILLEY, D. R. and MADANSKY, L. (1959) *Phys. Rev.* **116**, 413.
TILTON, G. R. (1951) *Atomic Energy Commission Report*, A.E.C.D.-3812.
TILTON, G. R. (1956) *Trans. Amer. Geophys. Union* **37**, 224.
TILTON, G. R. (1960) *J. Geophys. Res.* **65**, 2933.
TILTON, G. R., ALDRICH, L. T. and INGHRAM, M. G. (1954) *Anal. Chem.* **26**, 894.
TILTON, G. R., DAVIS, G. L., WETHERILL, G. W. and ALDRICH, L. T. (1957) *Trans. Amer. Geophys. Union* **38**, 360.
TILTON, G. R. and HART, S. R. (1963) *Science* **140**, 357.
TILTON, G. R., PATTERSON, C., BROWN, H., INGHRAM, M., HAYDEN, R., HESS, D. and LARSEN, E. S., Jr. (1955) *Bull. Geol. Soc. Amer.* **66**, 1131.
TILTON, G. R. and STEIGER, R. H. (1965) *Science* **150**, 1805.
TURNER, G., MILLER, J. A. and GRASTY, R. L. (1966) *Earth Planet. Sci. Letters* **1**, 155.

ULRYCH, T. J. (1967) *Science* **158**, 252.
ULRYCH, T. J. and REYNOLDS, P. H. (1966) *J. Geophys. Res.* **71**, 3089.
ULRYCH, T. J. and RUSSELL, R. D. (1964) *Geochim. Cosmochim. Acta* **28**, 455.
UMBGROVE, J. H. F. (1947) *The Pulse of the Earth*, Nijhoff, den Haag.
UREY, H. C. (1952) *The Planets*, Yale University Press.
UREY, H. C. (1964) *Revs. of Geophys.* **2**, 1.

VENING MEINESZ, F. A. (1951) *Trans. Amer. Geophys. Union* **32**, 531.
VENING MEINESZ, F. A. (1952) *Proc. Royal Neth. Acad. Sci.* **55**, 527.
VINCENT, E. A. (1960) in *Methods in Geochemistry*, ed. SMALES, A. A. and WAGER, L. R., Interscience London and New York, p. 148.
VINE, F. J. (1966) *Science* **154**, 1405.
VINE, F. J. and MATTHEWS, D. H. (1963) *Nature* **199**, 947.
VINE, F. J. and WILSON, J. T. (1965) *Science* **150**, 485.
VINOGRADOV, A. P. and TUGARINOV, A. E. (1961) *Ann. N.Y. Acad. Sci.* **91**, 500.
VOITKEVICH, G. H. (1958) *Priroda* **77**.
VON WEISZÄCKER, C. F. (1937) *Phys. Z.* **38**, 5, 623.

WAGER, L. R. (1964) *Quart. J. Geol. Soc. Lond.* **120S**, 13.
WÄNKE, H. and KÖNIG, H. (1959) *Z. Naturforsch.* **14a**, 860.
WANLESS, R. K. and LOWDEN, J. A. (1963) *Age Determinations and Geological Studies*, Geol. Surv. Can., Dept. Mines Tech. Surv. Paper 62–17, 121.
WASSERBURG, G. J. (1954) in *Nuclear Geology*, Wiley, New York, p. 341.
WASSERBURG, G. J. (1961) *Ann. N.Y. Acad. Sci.* **91**, 583.
WASSERBURG, G. J. (1963) *J. Geophys. Res.* **68**, 4823.
WASSERBURG, G. J. (1966) in *Advances in Earth Science*, M.I.T. Press, Cambridge, p. 431.
WASSERBURG, G. J., BURNETT, D. S. and FRONDEL, C. (1965) *Science* **150**, 1814.
WASSERBURG, G. J., FOWLER, W. A. and HOYLE, F. (1960) *Phys. Rev. Letters* **4**, 112.
WASSERBURG, G. J. and HAYDEN, R. J. (1955) *Geochim. Cosmochim. Acta* **7**, 51.
WASSERBURG, G. J., HAYDEN, R. J. and JENSEN, K. J. (1956) *Geochim. Cosmochim. Acta* **10**, 153.
WASSERBURG, G. J., WETHERILL, G. W., SILVER, L. T. and FLAWN, P. T. (1962) *J. Geophys. Res.* **67**, 4021.
WEBB, A. W. and MCDOUGALL, I. (1964) *J. Geol. Soc. Australia* **11**, 151.
WEBB, A. W. and MCDOUGALL, I. (1967) *Earth Planet. Sci. Letters* **2**, 483.

WEBB, A. W., McDOUGALL, I. and COOPER, J. A. (1963) *Nature* **199,** 270.

WEBSTER, R. K., MORGAN, J. W. and SMALES, A. A. (1957) *Trans. Amer. Geophys. Union* **38,** 543.

WESTCOTT, M. R. (1966) *Nature* **210,** 83.

WETHERILL, G. W. (1956) *Trans. Amer. Geophys. Union* **37,** 320.

WETHERILL, G. W. (1957) *Science* **126,** 545.

WETHERILL, G. W., ALDRICH, L. T. and DAVIS, G. L. (1955) *Geochim. Cosmochim. Acta* **8,** 171.

WHITE, F. A., COLLINS, T. L. and ROURKE, F. M. (1956) *Phys. Rev.* **101,** 1786.

WHITE, W. H., ERICKSON, G. P., NORTHCOTE, K. E., DIROM, G. E. and HARAKAL, J. E. (1967) *Can. J. Earth Sci.* **4,** 677.

WICKMAN, F. E. (1942) *Geol. Foren. Stockholm Forh.* **64,** 465.

WILSON, J. TUZO (1951) *Pap. and Proc. Roy. Soc. Tasmania* **85.**

WOOD, J. A. (1963) *The Moon, Meteorites and Comets, The Solar System,* Vol. 4, Univ. of Chicago Press.

WÜRGER, E., MEYER, K. P. and HUBER, P. (1957) *Helv. Phys. Acta* **30,** 157.

YORK, D. (1966) *Can. J. Phys.* **44,** 1079.

YORK, D. (1967) *Earth Planet. Sci. Letters* **2,** 479.

YORK, D. (1969) *Earth Planet. Sci. Letters* **5,** 320.

YORK, D., BAKSI, A. K. and AUMENTO, E. (1969) *Trans. Amer. Geophys. Union* **50,** 353.

YORK, D., McINTYRE, R. M. and GITTINS, J. (1969) *Earth Planet. Sci. Letters* **7,** 25.

ZARTMAN, R. E. (1964) *J. Petrol.* **5,** 359.

ZARTMAN, R. E. (1965) *J. Geophys. Res.* **70,** 965.

INDEX